AF388948

Low Energy Architecture and Low Carbon Cities

Low Energy Architecture and Low Carbon Cities: Exploring Links, Scales, and Environmental Impacts

Editors

Francesco Pomponi
Bernardino D'Amico

MDPI • Basel • Beijing • Wuhan • Barcelona • Belgrade • Manchester • Tokyo • Cluj • Tianjin

Editors
Francesco Pomponi
Edinburgh Napier University
UK

Bernardino D'Amico
Edinburgh Napier University
UK

Editorial Office
MDPI
St. Alban-Anlage 66
4052 Basel, Switzerland

This is a reprint of articles from the Special Issue published online in the open access journal *Sustainability* (ISSN 2071-1050) (available at: https://www.mdpi.com/journal/sustainability/special_issues/Architecture_Low_Carbon).

For citation purposes, cite each article independently as indicated on the article page online and as indicated below:

LastName, A.A.; LastName, B.B.; LastName, C.C. Article Title. *Journal Name* **Year**, *Volume Number*, Page Range.

ISBN 978-3-03943-815-0 (Hbk)
ISBN 978-3-03943-816-7 (PDF)

Cover image courtesy of unsplash.com user Terence Starkey.

Contents

About the Editors

Francesco Pomponi is an Associate Professor of Sustainability Science at Edinburgh Napier University, where he leads the Resource Efficient Built Environment Lab (REBEL), and Academic Tutor at the University of Cambridge's Institute for Sustainability Leadership. After six years in industry, he transitioned to academia to work on advancing global sustainability with a primary focus on the environmental impacts caused by buildings and construction. His award-winning research has been funded by the UK's Engineering and Physical Sciences Research Council, the Royal Academy of Engineering and the European Union, as well as governments and not-for-profit organizations. He is a UK member for Annex 72 of the International Energy Agency (IEA) and serves as a reviewer and expert advisor to the IEA Policy Division. With multiple editorial appointments in premiere journals, Francesco takes part in global collaborations with leading international scholars from Australia to the United States.

Bernardino D'Amico is an Associate Professor of Construction at Edinburgh Napier University, and a professionally qualified architect. After a few years in industry working as an architectural engineer on various research-driven projects and several research collaborations, in October 2012, he moved to academia, earning a PhD in computer-aided methods for the design of free-form grid-shell structural systems. His research interests lie in the general realm of design, analysis and fabrication of sustainable building structures, with a focus on computational structural design and optimization. In February 2017, he co-founded, with his colleague Dr Francesco Pomponi, the REBEL (Resource Efficient Built Environment Lab) Group, which aims to address sustainability issues in the built environment holistically.

Editorial

Low Energy Architecture and Low Carbon Cities: Exploring Links, Scales, and Environmental Impacts

Francesco Pomponi [1,2,*] **and Bernardino D'Amico** [1]

1   Resource Efficient Built Environment Lab (REBEL), Edinburgh Napier University, Colinton Road, Edinburgh EH10 5DT, UK; b.damico@napier.ac.uk
2   Cambridge Institute for Sustainability Leadership (CISL), University of Cambridge, 1 Trumpington Street, Cambridge CB2 1QA, UK
*   Correspondence: f.pomponi@napier.ac.uk

Received: 2 November 2020; Accepted: 4 November 2020; Published: 5 November 2020

Abstract: Projected population growth and urbanization rates will create a huge demand for new buildings and put an unprecedented pressure on the natural environment and its limited resources. Architectural design has often focused on passive or low-energy approaches to reduce the energy consumption of buildings but it is evident that a more holistic, whole-life based mindset is imperative. On another scale, the movement for, and global initiatives around, low carbon cities promise to deliver the built environment of tomorrow, in harmony with the natural boundary of our planet, the societal needs of its human habitants, and the required growth for economic prosperity. However, cities are made up of individual buildings and this intimate relationship is often poorly understood and under-researched. This multi-scale problem (materials, buildings, and cities) requires plural, trans-disciplinary, and creative ways to develop a range of viable solutions. The unknown about our built environment is vast: the articles in this special issue aim to contribute to the ongoing global efforts to ensure our built environments will be fit for the challenges of our time.

Keywords: low carbon cities; low energy buildings; sustainability transitions; shelter; building stock; building lifetime; carbon flux

---

1. Introduction

Projected population growth and urbanization rates will create a huge demand for new buildings [1,2] and put an unprecedented pressure on the natural environment and its limited resources [3,4]. Architectural design has often focused on passive or low-energy approaches to reduce the energy consumption of buildings [5] but it is evident that a more holistic, whole-life based mindset is imperative [6]. On another scale, the movement for, and global initiatives around, low carbon cities promise to deliver the built environment of tomorrow, in harmony with the natural boundary of our planet, the societal needs of its human habitants, and the required growth for economic prosperity. However, cities are made up of individual buildings and this intimate relationship is often poorly understood and under-researched (e.g., [7,8]). UN Secretary-General Ban Ki-moon said "Our struggle for global sustainability will be won or lost in cities." [9]; the scale of the challenge is enormous and at stake is our planet and the very livability of a human existence as we know of it, and as we imagine and hope our lives could, and should, be. Truly understanding the problem is intimately interwoven with a sense of urgency and a great worry. We hope that the contributions in this special issue can go some way towards mitigating the environmental impacts of buildings and cities, advancing the still very limited understanding of our own built environments, and contributing to the global movements for regenerative approaches to the design of buildings and cities in order to protect the life we have on the planet and, if possible, restore the life we have lost.

2. Context

The sustainability of buildings and cities is a long-standing, contentious issue. The movement of the Garden City, for instance, started at the beginning of the previous century and clearly clashed with the over high-rise construction in many parts of the world in the 1950s due to housing shortage [10]. Towards the end of the 20th century, technology and computer modelling pushed strongly towards a positivist paradigm of techno-optimism led to believing the problem had been sorted [11]: we simply needed computer models (and their continuous refinement) in order to simulate, measure, and improve everything. An unrealistic techno-optimism has also been identified as a cause to lack of progress on resource efficiency [12]. This lasts to these very days, with the next big step seemingly being a full integration between Building Information Modelling (BIM) and Geographic Information System (GIS) "for the future development of society, especially in the field of the sustainable built environment" (p. 50) [13]. However, it took nearly 400 years to advance—with pretty simple algebra—Galileo Galilei's square-cube law [14] to discover a dimensionless factor that consistently quantifies the degree of compactness of building forms only as a function of their shape, thus regardless of their size (and volume) [15].

On another scale, the circular economy has emerged as a wholly alternative paradigm to mainstream growth policies. The idea is simple: decoupling economic growth from environmental impacts, but with its 114 definitions (and counting ...) [16] a widely agreed understanding of what it actually is seems to still be lacking. Also, its materialization is stagnant: while there are significant advances in many sectors, these are often misrepresented or misinterpreted applications of the 3R principle (reduce, reuse, recycle) and the waste hierarchy. The complexity of the built environment further hinders a straightforward applicability of the concept of the circular economy [17], and while it is clear that technological and regulatory development will be needed [18], it is equally clear they will not suffice and need to be complemented by shifts in business models and people's behaviors [19].

Another barrier to a comprehensive understanding and advancement of the sustainability discourse in the built environment is the recurring, and wholly unnecessary, division between embodied and operational energy, CO_2 emissions, and—in general—environmental impacts [20]. Previous beliefs assumed operational energy would be all that matters since buildings have long lifespans. However, with increased energy efficiency and thus a reduction in operational energy demand, the balance is shifting, with embodied carbon quickly rising to dominate the global agenda on how to reduce the climate change impacts of buildings [21,22].

Lastly, a Global North-centric approach may also limit significant sustainability breakthroughs when 7 in 10 urban residents of our urbanized world live in developing countries, with the greatest urban population growth and urbanization expected to take place in Africa, Asia, and Latin America [9]. To this, one should add the increasing trend of worrying figures on global displacement due to natural disasters and human caused conflicts: there were 79.5 million forcibly displaced people worldwide at the end of 2019 (1% of the global population). Approximately 68% of them came from just five countries in the Global South and 73% are hosted by neighboring, likely developing, countries [23]. This South-South migration is often dominated by an urgency that impedes consideration of sustainability, despite refugee camps that are as big as medium-sized cities. As a consequence, our understanding of what actually matters in such contexts from an interdisciplinary sustainability perspective has only just began [24].

3. Content

It would be impossible for any collection of studies to comprehensively address the breadth, depth, and overall research size of the issues presented above. However, the articles in this special issue cover well several key elements that are crucial to foster our understanding of low energy architecture and low carbon cities.

Wang et al. [25] translate China's emission reduction regulations into a contextualized and tailored approach for a specific building type: high-speed railway station buildings in cold climates.

They reconcile the embodied and operational energy and emissions, through life cycle assessment (LCA) which is used in combination with BIM. Their analysis concludes that the ratio between embodied and operational impacts is 1:4, and they identify mitigation strategies that can address both. Space optimization for instance, reduces operational greenhouse gas (GHG) emissions as well as embodied emissions through reduced material demand. The highest reduction strategy, which combines interventions on space, envelope, and materials, shows a 28.2% improvement compared to a baseline scenario, and they also conclude on the important point of implementing measures that focus on space optimization early in the building design stage.

Zhang et al. [26] focus on the multi-scale carbon-carbon dioxide (C-CO$_2$) dynamics of subtropical urban forests and other green and grey infrastructure in Shanghai, China. Their work measures the C-CO$_2$ flux from different contributing areas depending on wind direction and atmospheric stability. Although the urban landscape they assessed is a net carbon source, urban forest patches functioned as a carbon sink. Their results aim to establish low-carbon-emitting planning and planting designs in the subtropics. Given China's rapid urban development, embedding green infrastructure in urban planning can represent a significant step forward towards the partial mitigation of impacts caused by urban agglomerates.

China is also the focus of the article by Zhou et al. [27]. They analyze building lifetime and stock turnover as the key determinants in modelling energy demand and emissions of building stocks. To address the data scarcity in official statistics or empirical data, they present a novel system dynamics model that adopts survival analysis to characterize the temporal differences between new construction, aging, and demolition of residential buildings in the Chinese context. The authors cover the entire residential stock of China over 11 years (2007–2017) and find that the average lifetime of urban residential building is around 34 years and that the overall stock size reached 23.7 billion m^2 in 2017. The implications of their results are profound if we reflect on the improvements that could be achieved by extending the service life to meet the full potential offered by materials and construction techniques, which generally goes well beyond a whole century.

China is a top contributor to both global energy demand and GHG emissions, and as such its low-carbon policies will have profound effects on us all. The effects of one such policies, the Low-Carbon Pilot Initiative (LCPI) that was implemented in July 2010, have been scrutinized in detail by Yu et al. [28]. Significantly, the authors moved from previous approaches based on estimates and worked on unified carbon emissions data for 1997–2015 obtained from the China Emission Accounts and Datasets. Their analysis employs a novel synthetic control method to establish the impact of LCPI on regional carbon emissions and uses the Guangdong Province—the largest of China's low-carbon pilot provinces—as a case study. Their results show a 10% abatement of emissions due to the implementation of LCPI, demonstrating the effectiveness of such policy but also warning on the need for continuous adjustments during the implementation.

A regional perspective also characterizes the work by Li et al. [29], who employed Social Network Analysis (SNA) to investigate the spatial correlation network structure of the CO$_2$ emissions in the Beijing–Tianjin–Hebei urban agglomeration. The authors construct a synergetic abatement effect model to calculate its effect in the cities and examine the influence that spatial network characteristics have on it. They find that Beijing and Tianjin are at the center of the emissions' spatial network, thus playing important roles that can control other carbon emission spillovers between the cities, obtaining a center-periphery structure of their network. This is useful in regional coordinated development, where areas with higher economic growth and lower pollution levels can be regarded as learning examples, and thus lead the way for other regions. Notably, their research shows that it is hard to achieve long-term emission reductions by solely imposing reduction targets in each individual city, whereas establishing a trans-regional emissions reduction coordination mechanism is suggested as the way forward. Similarly to Yu et al. [28], Li et al. [29] also call for a continuous adjustment and optimization of the spatial network structure of the CO$_2$ emissions.

Shi et al. [30] bring the focus back to cities and offer us a novel, standardized evaluation index system for low-carbon cities in China, applied to Xiamen as a case study. Notably, the authors integrate the perspective from thee index systems—(i) the Drivers, Pressures, State, Impact, Response model of intervention (DPSIR), (ii) a complex ecosystem, and (iii) a carbon source/sink process—to extract common indicators for low-carbon cities. They focus on quantitative input data to remove potential biases introduced by human subjective judgements. In the application to Xiamen as a case study over the 2010–2015 period, their analysis shows that the index of low-carbon development in 2015 was higher than that in 2010, while the rate of economic growth was greater than the growth rate of carbon emission, thus indicating that the relative decoupling of economic growth from carbon emission was to an extent partly achieved.

This concludes the works focused on China covered in our special issue. Such breadth of topics with some very novel contributions is refreshing given the sheer volume of impacts and building stock associated with China. The focus, spanning across different scales of analysis–from individual buildings to the trans-regional dimension of multiple cities–also demonstrates a much-needed multi-disciplinary approach to foster sustainable architectures and urban development.

The remainder of contributions in the special issue address three distinct topics, but the common theme is that none of them focuses on buildings or cities of well-known developed countries. We see this as a strength of the special issue, as seeking contributions from under-researched and underexplored areas has been one of our main objectives since the outset.

Abdullah and Alibaba [31] address the key topic of thermal comfort by focusing on window design and natural ventilation. This is already a crucial element in building design and will become even more important with a growingly warming global climate. Natural ventilation is a well-known passive design strategy that costs nothing but introduces the "indoor air quality-thermal comfort" dilemma, as the authors define it. Their analysis, which is based on computational modelling and simulation, addresses various degrees of window opening as well as different window-to-floor rations. In the Mediterranean climate of Cyprus, they find that unshaded windows under the most effective design and ventilation strategy are able to provide thermally comfortable indoor environment for 50–60% of the occupied hours. In addition to the factual finding, and in concordance with Wang et al. [25], they emphasize the importance of non-siloed sustainability considerations early in the design stage when room for improvement is at its maximum and the impact on costs minimal.

Reus-Netto et al. [32] offer a key contribution to mitigate the lack of utilization of energy rating systems in Latin America. They identify two issues: a lack of building energy efficiency regulations in some countries and an excessive complexity of such ratings or tools—where they do exist—that limits their adoption. To this end, they develop a simplified calculation method to estimate energy consumption of residential buildings. Their model is tested on 42 locations, by considering diverse climatic conditions and the fulfilment of different thermal transmittance requirements. The model offered in the article—which demonstrated a high reliability in a thorough statistical analysis presented by the authors—requires seven climatic characteristics as input data, and within the sociocultural context of Latin America has, according to the authors, more chance of being accepted and applied, thus increasing the rate of buildings with an energy assessment.

We conclude this editorial by presenting the only paper we co-authored in the special issue, in which Alshawawreh et al. [33] qualify the sustainability of novel designs and existing solutions for post-disaster and post-conflict (PDPC) sheltering. In this article the authors show that due to the constrained environment in which PDPC sheltering takes place, sustainability is seldom considered in spite of the severe practical consequences that doing so inflicts, both on people and the environment, as well as incurring higher costs in the long run. Significantly, Alshawawreh et al. [33] systematically categorize both existing solutions and novel designs along key (social, environmental, and economic) sustainability dimensions to identify best practices and learn lessons from what worked in the field and what did not. Each sustainability dimension is extensively discussed and analyzed and the article offers key recommendation for both the design stage and material choices. It is hoped that this work will represent a stepping stone to enable the growth of a new research area, that considers the sustainability of refugee shelters and camps as seriously and as imperatively as the sustainability of the buildings and cities that host us all.

Author Contributions: Conceptualization, F.P.; writing—original draft, F.P.; writing—reviewing and editing, F.P. and B.D. All authors have read and agreed to the published version of the manuscript.

Funding: The themes and research that allowed the idea and topic of this special issue to emerge and materialize have been funded by the UK's Engineering and Physical Sciences Research Council (EPSRC), grant number EP/R01468X/1, and the Royal Academy of Engineering (RAEng), grant numbers FoESFt5\100015 and FF\1920\1\19.

Acknowledgments: We are grateful to the authors of the contributions to this special issue for their valuable research.

Conflicts of Interest: The authors declare no conflict of interest.

References

1. D'Amico, B.; Pomponi, F.; Hart, J. Global potential for material substitution in building construction: The case of cross laminated timber. *J. Clean. Prod.* **2021**, *279*, 123487. [CrossRef]
2. Güneralp, B.; Zhou, Y.; Ürge-Vorsatz, D.; Gupta, M.; Yu, S.; Patel, P.L.; Fragkias, M.; Li, X.; Seto, K.C. Global scenarios of urban density and its impacts on building energy use through 2050. *Proc. Natl. Acad. Sci. USA* **2017**, *114*, 8945–8950. [CrossRef] [PubMed]
3. Miller, S.A.; Horvath, H.A.; Monteiro, P.J.M. Impacts of booming concrete production on water resources worldwide. *Nat. Sustain.* **2018**, *1*, 69–76. [CrossRef]
4. Pomponi, F.; Hart, J.; Arehart, J.H.; D'Amico, B. Buildings as a Global Carbon Sink? A Reality Check on Feasibility Limits. *One Earth* **2020**, *3*, 157–161. [CrossRef]
5. Berardi, U. A cross-country comparison of the building energy consumptions and their trends. *Resour. Conserv. Recycl.* **2017**, *123*, 230–241. [CrossRef]
6. Pomponi, F.; Moncaster, A. Circular economy for the built environment: A research framework. *J. Clean. Prod.* **2017**, *143*, 710–718. [CrossRef]
7. Susca, T.; Pomponi, F. Heat island effects in urban life cycle assessment: Novel insights to include the effects of the urban heat island and UHI-mitigation measures in LCA for effective policy making. *J. Ind. Ecol.* **2020**, *24*, 410–423. [CrossRef]
8. Newman, P.W.G. Sustainability and cities: Extending the metabolism model. *Landsc. Urban Plan.* **1999**, *44*, 219–226. [CrossRef]
9. Ki-Moon, B. UN Secretary-General Ban Ki-Moon's Remarks to the High-Level Delegation of Mayors and Regional Authorities—SG/SM/14249-ENV/DEV/1276-HAB/217. 2012. Available online: https://www.un.org/press/en/2012/sgsm14249.doc.htm (accessed on 1 November 2020).
10. Lock, D. The propaganda of the built environment. *RSA J.* **1991**, *139*, 455–466.
11. Hillier, B.; Penn, A. Virtuous circles, building sciences and the science of buildings: Using computers to integrate product and process in the built environment. *Des. Stud.* **1994**, *15*, 332–365. [CrossRef]
12. Allwood, J.M. Unrealistic techno-optimism is holding back progress on resource efficiency. *Nat. Mater.* **2018**, *17*, 1050–1051. [CrossRef] [PubMed]
13. Wang, H.; Pan, Y.; Luo, X. Integration of BIM and GIS in sustainable built environment: A review and bibliometric analysis. *Autom. Constr.* **2019**, *103*, 41–52. [CrossRef]
14. Galilei, G. *Discorsi e Dimostrazioni Matematiche Intorno a due Nuove Scienze*; Elsevirii: Leiden, The Netherlands, 1638.
15. D'Amico, B.; Pomponi, F. A compactness measure of sustainable building forms. *R. Soc. Open Sci.* **2019**, *6*, 181265. [CrossRef]
16. Kirchherr, J.; Reike, D.; Hekkert, M. Conceptualizing the circular economy: An analysis of 114 definitions. *Resour. Conserv. Recycl.* **2017**, *127*, 221–232. [CrossRef]
17. Pomponi, F.; Moncaster, A. Briefing: BS 8001 and the built environment: A review and critique. *Proc. Inst. Civ. Eng. Eng. Sustain.* **2019**, *172*, 111–114. [CrossRef]
18. Mercader-Moyano, P.; Esquivias, P.M. Decarbonization and Circular Economy in the Sustainable Development and Renovation of Buildings and Neighbourhoods. *Sustainability* **2020**, *12*, 7914. [CrossRef]

19. Hart, J.; Adams, K.; Giesekam, J.; Tingley, D.D.; Pomponi, F. Barriers and drivers in a circular economy: The case of the built environment. *Procedia CIRP* **2019**, *80*, 619–624. [CrossRef]

20. Pomponi, F.; Crawford, R.; Stephan, A.; Hart, J.; D'Amico, B. The 'building paradox': Research on building-related environmental effects requires global visibility and attention. *Emerald Open Res.* **2020**, *2*, 50. [CrossRef]

21. De Wolf, C.; Pomponi, F.; Moncaster, A. Measuring embodied carbon dioxide equivalent of buildings: A review and critique of current industry practice. *Energy Build.* **2017**, *140*, 68–80. [CrossRef]

22. Giesekam, J.; Pomponi, F. Briefing: Embodied carbon dioxide assessment in buildings: Guidance and gaps. *Proc. ICE Eng. Sustain.* **2018**, *171*, 334–341. [CrossRef]

23. UNHCR. United Nations High Commissioner for Refugees—2019 Figures at Glance. Available online: https://www.unhcr.org/uk/figures-at-a-glance.html (accessed on 27 October 2020).

24. Pomponi, F.; Moghayedi, A.; Alshawawreh, L.; D'Amico, B.; Windapo, A. Sustainability of post-disaster and post-conflict sheltering in Africa: What matters? *Sustain. Prod. Consum.* **2019**, *20*, 140–150. [CrossRef]

25. Wang, N.; Satola, D.; Houlihan Wiberg, A.; Liu, C.; Gustavsen, A. Reduction Strategies for Greenhouse Gas Emissions from High-Speed Railway Station Buildings in a Cold Climate Zone of China. *Sustainability* **2020**, *12*, 1704. [CrossRef]

26. Zhang, K.; Gong, Y.; Escobedo, F.J.; Bracho, R.; Zhang, X.; Zhao, M. Measuring Multi-Scale Urban Forest Carbon Flux Dynamics Using an Integrated Eddy Covariance Technique. *Sustainability* **2019**, *11*, 4335. [CrossRef]

27. Zhou, W.; Moncaster, A.; Reiner, D.M.; Guthrie, P. Estimating Lifetimes and Stock Turnover Dynamics of Urban Residential Buildings in China. *Sustainability* **2019**, *11*, 3720. [CrossRef]

28. Yu, X.; Shen, M.; Wang, D.; Imwa, B.T. Does the Low-Carbon Pilot Initiative Reduce Carbon Emissions? Evidence from the Application of the Synthetic Control Method in Guangdong Province. *Sustainability* **2019**, *11*, 3979. [CrossRef]

29. Li, X.; Feng, D.; Li, J.; Zhang, Z. Research on the Spatial Network Characteristics and Synergetic Abatement Effect of the Carbon Emissions in Beijing–Tianjin–Hebei Urban Agglomeration. *Sustainability* **2019**, *11*, 1444. [CrossRef]

30. Shi, L.; Xiang, X.; Zhu, W.; Gao, L. Standardization of the Evaluation Index System for Low-Carbon Cities in China: A Case Study of Xiamen. *Sustainability* **2018**, *10*, 3751. [CrossRef]

31. Abdullah, H.K.; Alibaba, H.Z. Window Design of Naturally Ventilated Offices in the Mediterranean Climate in Terms of CO2 and Thermal Comfort Performance. *Sustainability* **2020**, *12*, 473. [CrossRef]

32. Reus-Netto, G.; Mercader-Moyano, P.; Czajkowski, J.D. Methodological Approach for the Development of a Simplified Residential Building Energy Estimation in Temperate Climate. *Sustainability* **2019**, *11*, 4040. [CrossRef]

33. Alshawawreh, L.; Pomponi, F.; D'Amico, B.; Snaddon, S.; Guthrie, P. Qualifying the Sustainability of Novel Designs and Existing Solutions for Post-Disaster and Post-Conflict Sheltering. *Sustainability* **2020**, *12*, 890. [CrossRef]

Publisher's Note: MDPI stays neutral with regard to jurisdictional claims in published maps and institutional affiliations.

 *sustainability*

Article

Methodological Approach for the Development of a Simplified Residential Building Energy Estimation in Temperate Climate

Gabriela Reus-Netto [1,2], **Pilar Mercader-Moyano** [2,*] **and Jorge D. Czajkowski** [1]

[1] Sustainable Architecture & Habitat Laboratory, Faculty of Architecture & Urbanism, National University of La Plata, Calle 47 #162, 1900 La Plata, Argentina
[2] Dpto de Construcciones Arquitectónicas I, Escuela Técnica Superior de Arquitectura, Universidad de Sevilla, Avenida Reina Mercedes 1, 41012 Seville, Spain
[*] Correspondence: pmm@us.es

Received: 12 June 2019; Accepted: 22 July 2019; Published: 26 July 2019

Abstract: Energy ratings and minimum requirements for thermal envelopes and heating and air conditioning systems emerged as tools to minimize energy consumption and greenhouse gas emissions, improve energy efficiency and promote greater transparency with regard to energy use in buildings. In Latin America, not all countries have building energy efficiency regulations, many of them are voluntary and more than 80% of the existing initiatives are simplified methods and are centered in energy demand analysis and the compliance of admissible values for different indicators. However, the application of these tools, even when simplified, is reduced. The main objective is the development of a simplified calculation method for the estimation of the energy consumption of multifamily housing buildings. To do this, an energy model was created based on the real use and occupation of a reference building, its thermal envelope and its thermal system's performance. This model was simulated for 42 locations, characterized by their climatic conditions, whilst also considering the thermal transmittance fulfilment. The correlation between energy consumption and the climatic conditions is the base of the proposed method. The input data are seven climatic characteristics. Due to the sociocultural context of Latin America, the proposed method is estimated to have more possible acceptance and applications than other more complex methods, increasing the rate of buildings with an energy assessment. The results have demonstrated a high reliability in the prediction of the statistical models created, as the determination coefficient (R2) is nearly 1 for cooling and heating consumption.

Keywords: method of simplified calculation; energy consumption of buildings; multifamily residential building; temperate climate; Latin America

---

1. Introduction

Energy labelling of buildings and minimum requirements for insulation, solar control and performance of Heating, Ventilation and Air Conditioning (HVAC) systems emerged as tools to minimize the energy consumption and greenhouse gas emissions, improve the energy efficiency and promote greater transparency with regard to energy use in buildings [1].

The energy consumption of buildings is established by means of the relation between the energy demand and the performance of the mechanical space conditioning systems. The energy demand varies according to climatic conditions, the characteristics of the thermal envelope and the operative conditions of the building (occupancy, household appliance usage habits, equipment and lighting).

The first energy efficiency of building regulations emerged in the 1950s, in France, Switzerland and Germany, due to the concerns of the government entities to decrease the high heating energy consumption of buildings [2]. Rating initiatives emerged in the 1990s in United Kingdom, Germany and Canada with the objective to assess the energy efficiency and CO_2 emissions from buildings [3–5]. Currently, both regulations and rating schemes are very developed in the majority of these countries [2].

Regions in temperate climates represent the greatest territorial extension, with the highest population density, and so have the highest energy consumption. These countries cover 57% of the world population and are responsible of 68% of world's total primary energy consumption [6].

The temperate climates (C according to the Köppen classification)—Figure 1—are defined as having an average temperature above 0 °C in their coldest month but below 18 °C. Regarding the summer heat, "a" indicates the warmest month's average temperature is above 22 °C while "b" indicates the warmest month averages below 22 °C, but with at least four months averaging above 10 °C; and "c" indicates less than four months averaging above 10 °C. Countries like Morocco, Italy, Spain, Portugal, France, United States, Australia, Chile, Canada, Argentina, Brazil, Turkey, Switzerland, Greece, India, China, Japan, Mexico are examples of countries with locations with temperate climates [7].

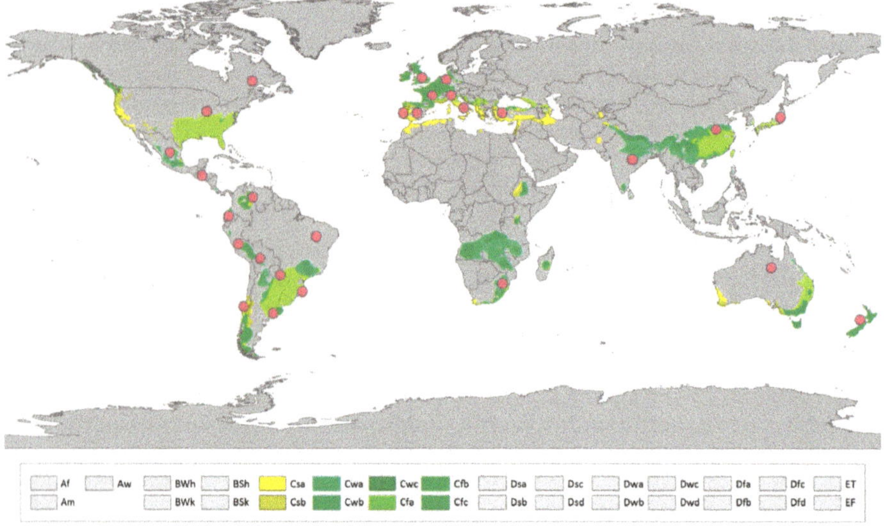

Figure 1. Development of building rating systems in temperate countries. Own elaboration adapted from the Köppen–Geiger climatic classification [7].

In order to know the energy rating context in countries with temperate climates, 47 initiatives for regulation and energy rating, distributed among 27 countries, were revised (Table 1). Almost half (48%) of the 47 initiatives revised are compulsory and cover the national territory. It has to be highlighted that, while all developed countries studies have energy efficiency regulations and energy rating schemes, not all emerging countries have the tools to assess the energy efficiency of their buildings.

Table 1. Building energy efficiency regulations and rating schemes in countries with temperate climates.

Country	Building Qualification System	Evaluation
	Africa	
Morocco	Thermal Regulation of Construction in Morocco—RTCM	C
South Africa	Energy efficiency in buildings—SANS 204	C
	Asia	
China	Evaluation standard for green building	C
China	Code for acceptance of energy efficient building construction	C
India	Energy conservation building code—ECBC	C
India	Green rating integrated habitat assessment—GRIHA	C
Japan	Design and Construction Guidelines on the Rationalization of Energy Use for Houses—Dcgreuh	R
Japan	Comprehensive Assessment System for Built Environment Efficiency—CASBEE	D, C
	Central America	
Costa Rica	Requirements for Sustainable Buildings in the Tropics—RESET	R
	Europe	
Germany	Passive house—Passivhaus	C
Germany	Energy Conservation Ordinance—EnEV	C
Spain	Technical building code—CTE	D
Spain	Energy certification of Spain	C
France	Energy performance diagnostic—DPE	C
France	Thermic regulation 2012—RT 2012	C
Italy	Decree 26 06 2015-Application of building energy performance calculation methodology and definition of minimum specifications and requirements for buildings (15A05198).	C
Portugal	Regulations on Thermal Behaviour of Buildings—RCCTE	D
Portugal	System for Energy and Indoor Air Quality Certification of Buildings	C
United Kingdom	Building Research Establishment Environmental Assessment Method—BREEAM	C
United Kingdom	Energy Performance Certificate—EPC	C
Swiss	Standard of thermal energy in building construction—SIA380/1	D
Swiss	Sustainable building standard—MINERGIE	C
Turkey	Thermal insulation requirements for buildings-TS 825	C
Turkey	Energy Performance Certificates	C
	North America	
Canada	Building Environmental Performance Assessment Criteria—BEPAC	C
Canada	Green Building Challenge—GBC	C
United States	Leadership in Energy & Environmental Design—LEED	C
United States	Building energy quotient—bEQ	C
Mexico	Sustainable Buildings Certification Program—PCES	C
Mexico	Mexican norm of sustainable building-NMX-AA-164-SCFI	D
	South America	
Argentina	Law 13059/03-Thermal Conditioning Conditions	R
Argentina	Energy performance in residential units—IRAM 11900	R
Argentina	Hygrothermal aspects and energy demand of buildings-Ordinance 8757/11	R
Brazil	Brazilian Building Labeling Program—PBE Edifica	C
Brazil	High environmental quality—AQUA BRAZIL	D
Brazil	Seal Blue House-SELO AZUL	R
Chile	Home Energy Rating—CEV	C
Chile	Sustainable Building Certification—CES	D
Colombia	Sustainable Construction Technical Regulation	R
Ecuador	Ecuadorian Construction Standard	R
Paraguay	Parameters of thermal comfort—NP 4901715	R
Peru	Thermal and Light Comfort with Energy Efficiency-EM.110	R
Uruguay	Reduction of energy demand for thermal conditioning-Resolution N° 2928/09	R
	Oceania	
Australia	Building Codes of Australia	C
Australia	Nationwide house energy rating scheme—NatHERS	C
New Zealand	New Zealand Building Code—NZBC	C

C = Energy consumption, D = Energy demand, R = fulfilment of normative requisites.

Due to the analysis of the application of these regulations and schemes, two contexts are highlighted: the European Union and Latin America.

In the European Union, policies and strategies of each country member are conducted by the European Directives, the directives 2002/91/CE [8] and 2010/31/UE [9] being the most influential in the implementation of energy efficiency regulations and of the building energy rating scheme. They require

European member states to achieve a particular result without dictating the means of achieving that result. Directives normally leave member states with a certain amount of leeway as to the exact rules to be adopted. Directives can be adopted by means of a variety of legislative procedures depending on their subject matter.

In Latin America, each country develops and implements the policies and strategies of its territory, as well as the regulatory framework for buildings. Not all Latin American countries have regulations focused on the energy assessment and energy efficiency of buildings. Furthermore, many regulations and energy rating schemes are voluntary [10].

The main compulsory initiatives in Latin America are the Brazilian Programme for Energy Labelling of Buildings (PBE Edifica) and the case of Rosario (Argentina). The first one requires an A level for new public buildings; the second one requires the fulfilment of certain values for thermal transmittance, risk of condensation, air permeability and cooling and heating thermal loads, established in the Ordinance 8757/11, in order to obtain the necessary licenses [10].

Regarding the assessed values of the building energy efficiency initiatives revised Table 1, in Latin America, 62% assess the achievement of certain requisites (thermal transmittance, solar factor, thermal capacity ...), 19% evaluate the energy demand and 19% focus on energy consumption. In Europe and the rest of the world, these proportions are different, with the main proportion of these initiatives centered on energy consumption, followed by the assessment of the energy demand, shown in Figure 2.

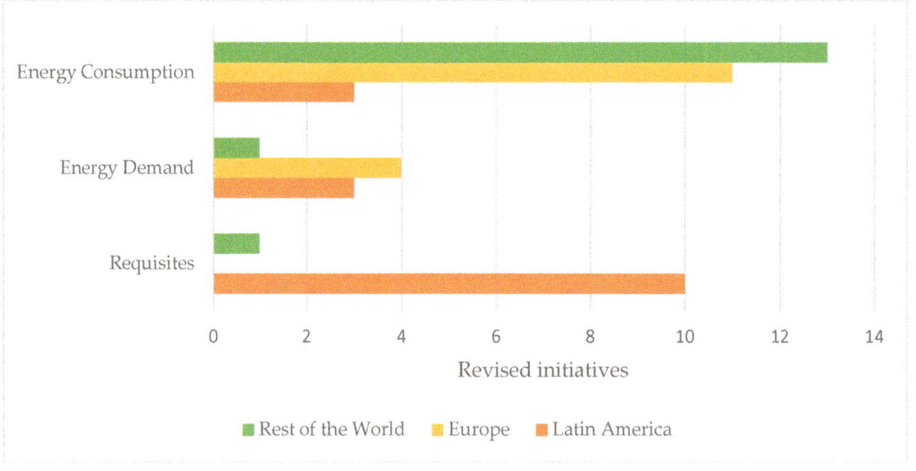

Figure 2. Evaluation criteria according to the regulations and certifications of countries with a temperate climate.

According to Krstić and Teni [11], building performance evaluation models can be classified as black, grey or white box. The black box is the model with the least information required to the user and the white box is the model with the highest information required to perform the energy evaluation of the building.

A black box model uses a simple mathematical or statistical model which relates a set of influential input parameters. The energy prediction under this model is based on a statistical approach according to a relevant database.

A white box model is based on building physics, and requires a detailed description of the building and its thermal systems. What is known as the general method, based on energy simulations using DOE 2 or EnergyPlus, which calculates the dynamic energy performance of the building [12], can be included in this classification. This model uses complex tools, requiring a high level of expertise, and consumes lots of time and resources [13].

A grey box model mixes both above-mentioned methods, as it requires certain key parameters identified from a physical model and the energy prediction is based on statistical methodologies. In these models, a rough description of the building geometry is enough.

The input data for simplified methods (black or grey box models) are diverse. The analysis of 43 simplified methods focused on energy consumption (Table 2) shows that the most common input data is weather data (16), followed by characteristics of the building (13), energy bills (13), occupancy loads (11), characteristics of the thermal envelope (8), characteristics of the HVAC systems (5), building typology (5) and other inputs. In total, 23 of those methods are destined to residential buildings, 14 for office buildings, 5 for commercial buildings and only one is for any use.

Table 2. Energy building performance simplified models: input data.

Year	Authors	Country	Typology	Input Data
1991	Bartels & Fiebig	Australia	R	HC
1994	LaFrance & Perron	Canada	R	W + EE + P
1995	Kreider et al.	United States	O	W + HC
1995	Hsiao et al.	-	R	O + HC
1996	Jaccard & Baille	Canada	R	HC
1998	Farahbakhsh et al.	Canada	R	B + O
1999	Fung et al.	Canada	R	B + W + EE + P + PE
2000	Kalogirou & Bojic	-	R	W + BE
2002	Shipley et al.	Canada	O	B + HC
2002	Lins et al.	Brazil	R	HC
2002	Mihalakakou et al.	Greece	R	W
2004	Shimoda et al.	Japan	R	O
2005	Parekh	-	A	B + O
2006	Petersdorff et al.	European Union	R	W + BE + T
2007	Kadian et al.	India	R	HC
2007	Raffio et al.	-	R	W + HC
2008	Swan et al.	Canada	R	B
2009	Hu	China	O	W + B
2010	Fumo	United States	R	HC
2010	Lam	China	C	B + O + EE
2010	Li	China	R	B
2010	Min et al.	United States	R	O + HC
2010	Wong et al.	-	O	W + BE + O
2011	Escriva-Escriva et al.	Spain	O	HC
2012	Melo	Spain	O	B + T
2012	Aranda	Brazil	C	B + W + EE
2013	Filippín	Argentina	R	HC + M
2013	Korolija	United Kingdom	O	B + T + EE + BE + O
2013	Zhou S. & Zhu N	China	O	W + BE
2014	Asadi	United States	C	BE + T
2014	Braun	United Kingdom	C	HC
2014	Fan	China	O	W
2014	Farzana	China	R	W + O
2014	Jain	United States	R	W
2014	Johnson	United States	R	O
2014	Mena	Spain	O	W
2014	Mastrucci	-	R	B + P
2015	Shams Amiri	United States	C	BE + T + O
2015	Salvetti	Argentina	R	B + O
2016	Pulido-Arcas	Chile	O	BE + EE
2017	Pino-Mejías	Chile	O	BE
2018	Nath Lopes & Lamberts	Brazil	O	W + B + BE + EE + O
2018	Ran Yoon	South Korea	O	O

Typology: R= residential, O = office, C = commercial, A = any type. W = weather, HC = historical consumption, B = building, BE = Building Envelope, EE = Equipment Efficiency, M = measured, O = occupation, P = demographic density, PE = price of electric power, T = typology.

The simplified method of Petersdoff, Boernans and Hamish [14] is based on five building typologies with eight different levels of thermal insulation simulated on three climatic regions of Europe: cold, moderate and warm. Wong, Wan and Lam [15] proposed a model for office buildings with daylighting controls in subtropical climates, considering four climatic variables, four envelope variables and the type of day. Korolija, et al. [16] developed a model to predict the energy consumption of HVAC in office buildings in the United Kingdom, from four building typologies, seven envelope variables, five HVAC systems and five operational variables. Nath and Lamberts [17] established a simplified model to estimate the cooling energy consumption of an office building by simulating it on 18 Brazilian cities. They modified 15 building parameters, 12 HVAC system variables and three occupancy variants. The cities were selected based on a classification of the Brazilian local climates-cities-on 18 groups based on an index built from degree-hours and enthalpy.

The presence of general methods in Latin America is very low and their application is very scarce. The lower amount of required information and the lower time consumption of the simplified methods have allowed many Latin American national initiatives to be based on simplified methods. In fact, more than 80% of the Latin American initiatives are simplified methods and are centered on energy demand limitation or in the establishment of minimum requirements of the thermal envelope. However, the application of these tools is reduced.

It has to be highlighted that the energy assessment based on the energy demand or the characteristics of the thermal envelope does not integrate all the present heat interchange processes and does not account for energy consumption and greenhouse emissions, as the performance of HVAC systems are not taken into account.

The main objective is to develop a simplified calculation model to estimate energy consumption in order to assess the environmental impact of the building stock and to guarantee a higher acceptance and application in Latin America, with the objective of increasing the number of buildings with energy evaluations. This work is focused on multifamily housing buildings due to their high representativeness, especially in cities with a population greater than 200,000 habitants [18].

2. Materials and Methods

In order to create a simplified energy consumption estimation model for residential buildings in temperate climates for Latin America, the next steps were followed, as shown in Figure 3:

1. Analysis of current energy building performance initiatives.
2. Data collection: case study, user energy performance and present environmental conditions.
3. Development and calibration of the energy model.
4. Simulation scenarios: locations and U-value thresholds and proposals for walls and roofs.
5. Energy consumption and linear regression.
6. Equations for energy consumption estimation.

The aim of the first step is the finding of the common parameters of the thermal envelope of the residential building energy performance initiatives for those countries of Latin America with cities located in temperate climates, in order to get the threshold values, which will conform the simulation scenarios.

The second step aims at gathering all necessary information needed to define and characterize the case study in order to build the energy model: the geometrical and constructive definition of the case study, the elaboration of the users' profile and the present environmental performance of the case study [19]. This case study, a residential building, is representative of this type of residence, as shown in the national office of statistics of each selected country.

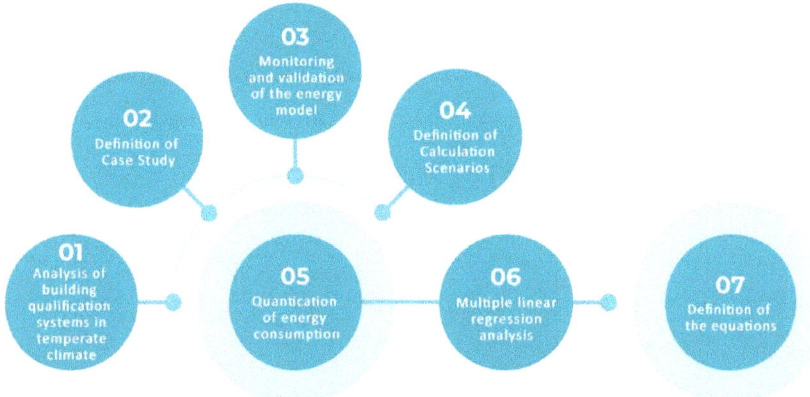

Figure 3. Scheme of the methodology created.

The third phase compares the monitored environmental performance of the case study with the simulated results in order to calibrate the energy model. Once the energy model was calibrated, the simulation scenarios were defined which are the function of the locations of the selected countries with temperate climates and the legislative threshold values for common thermal envelope parameters found in the first step. Finally, five constructive scenarios representing the variety of these threshold values are defined.

In order to get the energy consumption and its variation from the simulation scenarios, the thermal model of the original case study is simulated for each location of each country and for each location changing the constructive scenario according to its national requisites.

The correlation between the energy consumption data and the climatic variables is studied due to the development of linear regression statistical models where the most relevant variables are highlighted.

Finally, once these variables are identified, the correlational equations for heating and cooling estimation are defined, forming a simplified method for energy estimation based on few climatic data, easily available for building agents.

These steps can be grouped in two: development of the experiment and statistical analysis. The first group, covering the steps 1 to 4, is exposed in this chapter and the statistical analysis of the energy consumption and development of the correlational equations, steps 5 to 7, are shown in the results chapter.

2.1. Analysis of Current Energy Performance of Buildings Initiatives

Focused on Latin American context, the actual initiatives for energy performance of buildings centered on residential buildings were prioritized, which include objective requisites about comfort and energy efficiency. The initiatives for residential buildings from countries with at least five locations with temperate climate and the availability of weather files for EnergyPlus (epw) were selected.

The analysis also includes the Spanish scheme for being a referent in Latin America due to historical linkages and the use of a common language, among other factors. Furthermore, climatic similarities with the temperate regions of Latin America and their weather files for energy simulations are available, enabling a comparison between the Spanish minimum requisites and the Latin American threshold values [20–22].

Table 3 shows the National Building Energy Rating schemes, the mandatory standard that limits the characteristics of the thermal envelope and the parameters limited. In bold are the common parameters of the thermal envelope of the selected schemes.

Table 3. National Building Energy Rating (NBER) schemes for residential buildings and their correlated thermal envelope standard (TES) in Latin America and Spain.

Country	NBER and TES	Parameters of the Thermal Envelope
Argentina	IRAM 11900:2007 Law 13059:2003 IRAM 11604:2001 IRAM 11605:1996 IRAM 11625:2000 IRAM 11507-4:2010	**U-value: walls, roof**, floor, glazing Global losses coefficient Solar Heat Gain Coefficient (SHGC) Air infiltration rate Condensation risk
Brazil	PBE Edifica RTQ-R	**U-value: walls, roof** Thermal capacity: wall and roof Solar absorptivity for opaque enclosures Window to Wall ratio Natural ventilation factor
Chile	CEV Thermal Regulation	**U-value: walls, roof**, floor, glazing Thermal inertia SHGC Air infiltration rate
Mexico	Ecocasa NOM-020-ENER-2011 NOM-024-ENER-2012	**U-value: walls, roof**, floor, glazing Comfort range
Spain	RD 47/2007 RD 235/2013 Technical Building Code: Energy Saving Document (CTE DB-HE)	**U-value: walls, roof**, floor, glazing, internal partitions SHGC Air infiltration rate Condensation risk

The selected Latin American initiatives for energy efficiency of buildings were the Argentinian Thermal Conditioning Standards (Law 13059:2003) [23], the Brazilian Building Labelling Programme (PBE Edifica) [24], the Chilean scheme for Housing Energy Rating (CEV) [25] and the Mexican Programme Ecocasa [26].

In the Argentinian context there is an energy rating scheme for residential buildings, the IRAM 11900:2017 standard, but this is of voluntary fulfilment and is based on the requirements of the Law 13059:2003 which forces different IRAM standards related to the characteristics of the thermal envelope to be applied.

The Brazilian Programme includes the Technical Regulation of Quality for the Energy Efficiency Level of Residential Buildings (RTQ-R [27]), which limits the values of the different parameters of the thermal envelope according to the bioclimatic zone of the location. The Chilean context is similar; the Building Energy Efficiency Rating is based on the limited values determined in a specific legislative document called Thermal Regulation [28]. The Mexican Building Energy Rating is mandatory for new housing and is based on the calculation program of PassivHaus, which requires the minimum levels specified in the Official Mexican Standards (NOM), also compulsory. The Spanish context is similar to the Brazilian, Chilean and Mexican contexts: there is a Building Energy Rating scheme which also allows for the certification of the fulfilment of the thermal requirements of the thermal envelope.

The common parameter of the thermal envelope that is limited by the selected schemes is the thermal transmittance (U-value) of the external walls (vertical and horizontal, i.e., walls and roofs). So, this parameter is selected to be modified and then to build different simulation scenarios.

2.2. Data Collection

2.2.1. Case Study

Nowadays, more than 80% of the population of Latin America and the Caribbean live in cities [29]. Data from the Economic Commission for Latin America and the Caribbean (ECLAC) [30] show that the percentage of the population that lives in cities with more than 20,000 inhabitants in Argentina,

Brazil, Chile and Mexico is 79.8%, 70.6%, 80.3% and 70.2% respectively. Furthermore, according to the Economic Commission for Latin America and the Caribbean (CEPALSTAT) [31], the percentage of urban housing in these countries is 90%, 83%, 86% and 76% respectively, with approximately 75% on property.

Housing buildings are the most representative building typology in countries with temperate climates and are the greatest energy consumers during their useful life [18]. In Spain, 68% of residential units are located in multi-family buildings, being the majority in cities with more than 5000 inhabitants. It was found that 56.3% of residential units have no thermal insulation [32]. This percentage is lower in Latin America, but still of relevance; 57% of the Chilean residential buildings are located in metropolitan cities; of this buildings, 81% are multi-family residential buildings [32]. In Argentina, 21.38% of residential units are located in multi-family buildings in cities with more than 20,000 inhabitants [33]. In Brazil, the average amount of multi-family buildings is 35% [34]. A lower presence of multi-family buildings can be found in Mexico [35].

The case study is a multi-family building based on traditional building fabric, designated to middle class families according to the Brazilian Programme "Minha Casa Minha Vida" Figure 4.

Figure 4. Case study: multifamily building.

The structure is of reinforced concrete; the external walls are based on brick masonry with a plaster layer on both sides; the windows are simple glazed with aluminum frame without solar protection; the roof is based on a concrete slab and aluminum–zinc tiles. The U-values are 2.51 W/m^{2}K for the external walls, 1.96 W/m^{2}K for roofs, 5.80 W/m^{2}K for windows and 2.88 W/m^{2}K for the internal floors.

The building is located in a plot of land at the center of Criciúma, Santa Catarina, Brazil, being 28°41′ South its latitude. According to Köppen classification, its climate is Cfa (humid subtropical climate), characterized by hot and humid summers, and mild winters [7]. According to the Brazilian Standard NBR 15.220: Thermal Performance of Buildings [36], Criciúma is located in climatic zone II, characterized by an annual average daily temperature higher than 18 °C, as shown in Figure 5. The urban land is mainly plane, the surrounding is based on isolated multi-family buildings, so there is no vegetation or other buildings to obstruct the solar incident radiation or exposure to wind.

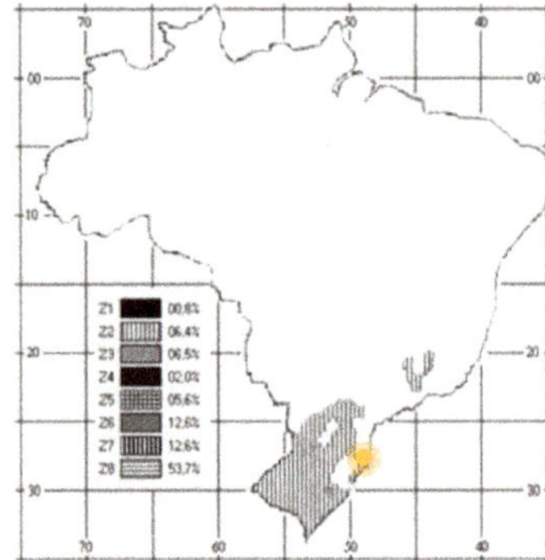

Figure 5. Location of Criciúma: zone II of the bioclimatic map of Brazil.

2.2.2. User Energy Performance

In order to define the occupancy and usage profile to be incorporated to the energy model, a survey of the inhabitants of the case study was carried out. A questionnaire was developed as a google formulary and sent by email to the responsible of each residential unit. From a population of 480 inhabitants, a response rate of 11% was recorded.

The focus of the survey was the understanding of those phenomena that could influence perceived comfort levels by recollecting information related to occupancy habits, HVAC preferences, possible modification of the internal covering, as well as perceptions about thermal comfort inside their dwellings.

The structure of the survey was composed of two introductory sections and four sections asking about occupancy, activities, timing, thermal comfort perception, air conditioning equipment and actions followed to reach thermal comfort Figure 6.

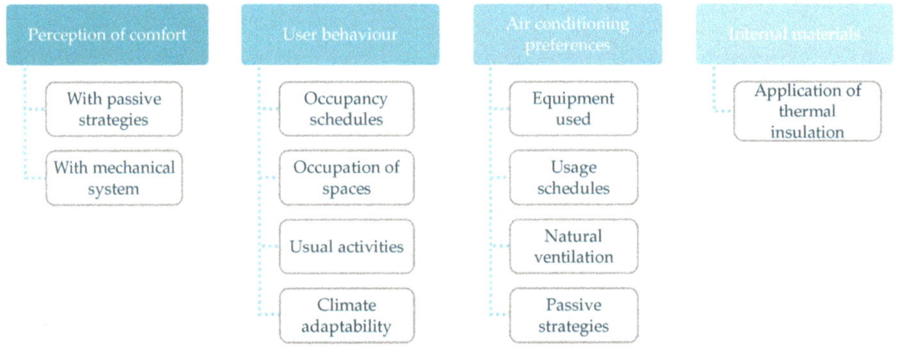

Figure 6. Fields of observation.

The introductory sections were informative and exposed the objective of the survey, its extent and the expected time to fulfil it. The first group of questions was focused on the identification of the users' opinion about their perceived thermal comfort under two situations: using or not using HVAC equipment. The following section asked about their habits: how they occupy and use their dwelling and how they are dressed while they stay at home. The next section focused on the preferences for using HVAC equipment: type of HVAC equipment, frequency of use, operation mode, natural ventilation and other passive strategies followed by the users. The last section asked about the constructive state of the dwelling in order to know if the thermal envelope had been modified.

The survey was based on multiple choice questions. An example can be found in Figure 7. Once the answers were received, they were analyzed in order to get statistical information about user behavior (an example is shown in Figure 8). Based on this statistical information, the user profile could be created.

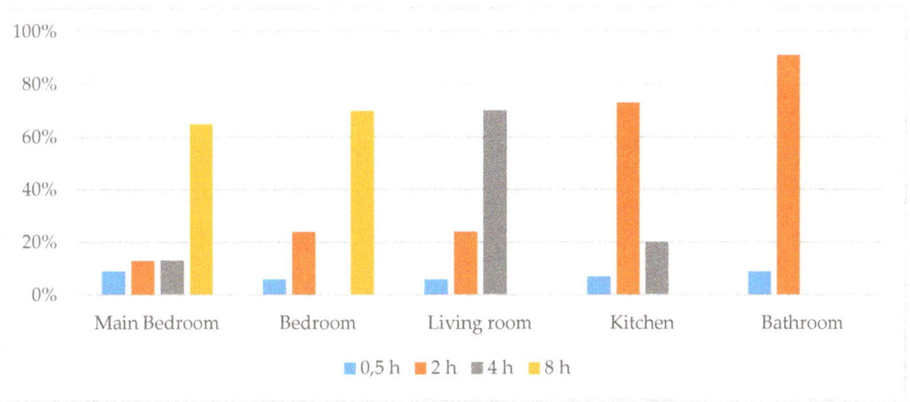

Figure 7. Multiple choice question example.

Figure 8. Statistical analysis of one of the questions of the survey.

The surveys showed that two-bedroom dwellings were mostly occupied by two inhabitants; meanwhile four inhabitants were the majority occupancy for three-bedroom dwellings. The daily time duration of each user activity was 8 h for sleeping, 8 h for working, 2 h for cooking and eating, 2 h for cleaning and personal care and 4 h for entertainment.

The users adapted their clothing according to seasonal variations, being tropical for summer (clo = 0.3), intermediate for spring and autumn (clo = 0.5) and light coated for winter (clo = 1.0).

Bathrooms, kitchen and common circulation areas were spaces where ventilation was the main strategy for thermal conditioning, meanwhile bedrooms and the living room used air conditioning equipment (if any). The HVAC systems on dwellings are air-to-air heat pump splits, fans and radiators. During summer, 100% of inhabitants used fans to reach thermal comfort at any time during the day, meanwhile 63% of the population use the split heat pump. However, 63% opened the window every time they wanted to ventilate the dwelling. Only 16% of the users maintained closed windows. During spring and autumn, 46% used fans, 9% used the split and 4% needed to use electric radiators. In these seasons there was a great portion (37%) which did not use any mechanical space conditioning system, and 24% kept their windows closed. During winter, those occupants with split systems used them (63%), 22% had and used electric radiators and 20% of the population did not use any heating equipment or did not have it. In this season, 54% of the inhabitants kept their windows closed.

The dwellings are equipped with a fridge (24 h functioning), two PCs and two TVs that are utilized for 2 h each, and the shower is used for 1 h. Daylight is the main lighting source from 8:00 h to 18:00 h and there is a fluorescent lamp in each space which functions from 18:00 h to 22:00 h.

Surveys verified that inhabitants used adaptive comfort practices. The operation hypothesis, thus, has been assembled bearing in mind the strategies and actions that are allowed per adaptive comfort model, that is, the changes of clothing for inhabitants and the operation of windows in order to get the dwelling ventilated, shown in Table 4 and Figure 9.

Table 4. Operation hypothesis.

	Main Bedroom	Bedroom	Living Room	Kitchen	Bathroom
People	2	1	2	1	1
Hours	8	8	4	2	2
Activity	Sleep	Sleep	Read/eat	Cook	Shower
Metabolism (W/pers)	72	72	110	230	180
Clothing (clo)	Summer = 0.3	Winter = 1,0	Spring-Autumn = 0.5			
Thermal zone	Conditioned			Ventilated	
Tª setpoint	Minimum = 18 °C	Maximum = 27 °C		-	
Air change rate	2				

Figure 9. Occupancy hypothesis.

The environmental temperature range for comfort and HVAC was taken from the adapted Givoni's chart, with a minimum value of 18 °C and a maximum of 27 °C.

The air change rate was set to 2 ac/h due to the ventilation strategy of the users and the bad quality of permeability of the window frames. This value is in accordance with the suggested values of the standard IRAM 11.604/01 [37].

2.2.3. Present Environmental Conditions

In order to get the present environmental conditions of the building for the purpose of calibrating the energy model, seven dwellings were monitored. Dry bulb air temperature (inside–outside), relative humidity (inside–outside) and global exterior solar radiation were recorded for a time-step of 30 min from the 14th to 22th in January and from 10th to 18th in July, representing winter and summer conditions, both in 2016.

Inside thermal conditions were recorded by two HOBO UX100-003 data loggers per dwelling (in the main bedroom and living room), located 150 cm from the floor and protected from direct solar radiation, shown in Figure 10. The external temperature, relative humidity and global radiation were recorded with another HOBO data logger. The sensors were protected from direct solar radiation and precipitation.

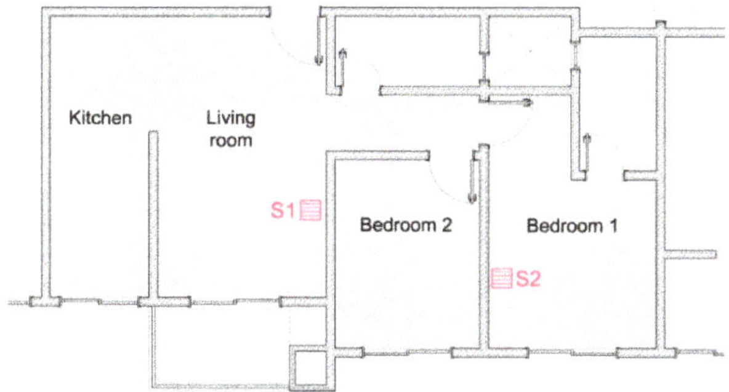

Figure 10. Location of the sensors in ground plan: Apartment 303.

The measured values were drawn in the adapted Givoni's chart, as it takes into account the adaptability of the user to different temperatures and relative humidity ranges [38]. Although it is designed to include only outdoor air conditions, indoor measured values were drawn in order to confirm the strategies followed by users, as they had reflected on the surveys. The first comfort range (zone E) considers relative humidity values varying from 30% to 50%; the second range (zone F) considers an adaptation to relative humidity values of up to 80%. The temperature for a comfort zone with low humidity varies from 18.5 °C to 27.5 °C but varies from 17 °C to 26 °C if the relative humidity is high, shown in Figure 11.

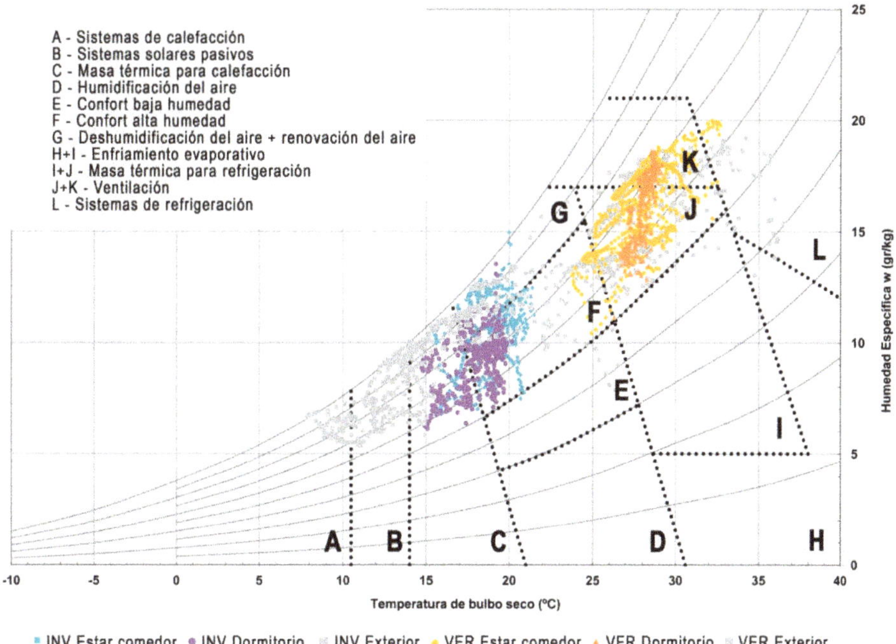

Figure 11. Measured temperature and relative humidity in the adapted Givoni's chart.

It can be observed that during the summer week, 3.24% of the measured hours in the living room were in thermal comfort zone E without any air conditioning strategy. Considering the natural ventilation and adaptation to high humidity, the percentage of hours in comfort increased to 28.70%.

In the main bedroom, those values were 0.46% and 10.19%, respectively. It can be seen that the main bioclimatic strategy followed by users to reach the comfort zone during summer is ventilation, as air movement reduces the perceived temperature.

During the winter week, 6.49% of the measured hours in the living room were in comfort zone E without passive strategies. As users in this period use artificial heating systems, and let solar radiation enter the dwelling but keep the windows closed, the humidity increases but within a certain range, and comfort hours increased this percentage to 49.80%. In the case of the main bedroom, those values were 4.24% and 38.89%, respectively. It can be observed how the users' strategies, obtained from the surveys, during winter have a higher impact on reaching the comfort zone, even with high humidity, than in the summer time.

2.3. Development and Calibration of the Energy Model

An energy model of the case study was created in DesignBuilder (version 5.0.0.137), based on the widely respected EnergyPlus simulation program [39]. The input climate data were the energy plus weather data (epw) [40], modified with the external measured data.

Geometric, constructive, occupancy, equipment, HVAC systems and usage characteristics were created according to the obtained data from the documentation of the building and the user profile created from surveys.

In order to validate the thermal model, a comparison between simulated internal temperature on free running and the measured values from monitoring Figure 12.

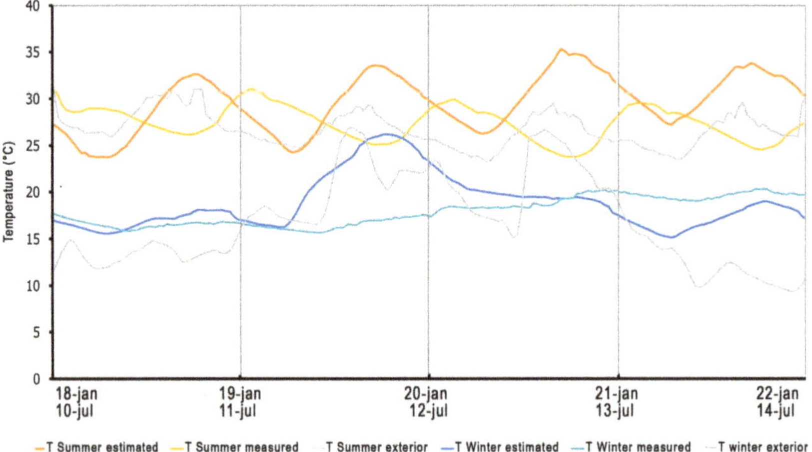

Figure 12. Comparison between measured and estimated interior temperature.

It can be observed that, in summer, the maximum measured temperature was 31.1 °C, meanwhile the maximum estimated temperature was 35.3 °C. In winter, the maximum inside measured temperature was 20.3 °C and the estimated was 26.2 °C with a thermal time lag of 11 h. These values indicate that the energy model overestimates the temperature values, so it was necessary to identify the sources of uncertainty in order to calibrate the model.

Regarding the measured values, it was observed that the heterogeneity of the values was the main uncertain source. The total typical uncertainty of the measured values was 0.15 °C (95% confidence interval, coverage factor k = 2) indicating that the uncertainty is low and that the air stratification during the most critical summer and winter week is negligible.

The geometry of the model was the actual geometry, as it was measured from the case study. The properties of the materials were manually introduced based on the real construction. The weather file is a statistical compilation, so there is a difference between measured data and the data contained in the weather files. Occupancy and internal loads were also a source of uncertainty, as the user profile was created based on the survey, which implies certain subjectivity.

A sensitivity analysis was carried out between three outputs and the parameters of thermo-physical properties of the materials, the general conditions of the building and its surrounding in order to assess the uncertainty of the energy model. The operative temperature was found to be the output which better reflects the modification of the parameters of the model. To adjust the model, the most influential parameters were the air change rate and the thermal conductivity of concrete in roofs and of the brick layer in walls.

The result of the energy balance showed that internal loads and solar heat gains were a significant heat source that provoked the estimated temperatures to be higher than the measured ones. Internal equipment loads and lighting were eliminated from the model and, due to the results from the sensitivity analysis, the air change rate was modified. Thermal properties of the roof and walls were not modified in order to keep the representativeness of the simulation scenarios.

Calibration of the model was performed according to guidelines from ASHRAE Guideline 14 [41]. This guideline states that a model could be considered validated if its mean bias error (MBE) is no larger than 10%, and if the coefficient of variation of the root-mean-square error (CV(RMSE)) is not larger than 30% when the hourly data is used for the validation. The validation was based on the monitored and simulated dry bulb indoor air temperature, which was measured in 30-min intervals.

In each monitoring period, the MBE and CV(RSME) satisfied the 10% and 30% limits respectively for indoor air temperature, as recommended in ASHRAE Guideline 14.

2.4. Simulation Scenarios

In order to build a simplified calculation method and assess the energy consumption difference due to the fulfilment of the minimum requisites of thermal transmittance defined by the Argentinian, Brazilian, Chilean, Mexican and Spanish initiatives, two simulation series were carried out. The first one consists of a simulation of the case study built in different locations with temperate climates. The second one consists of a simulation of the case study in these locations but modifying the thermal transmittance of the opaque elements of the thermal envelope according with the legal requisites of the country.

2.4.1. Locations and U-value Thresholds

In order to establish the locations with temperate climates for the first set of simulations, cities from Argentina, Brazil, Chile, Mexico and Spain with temperate climates and a population higher than 200,000 inhabitants were selected, totaling 307 locations. The information about population was obtained from national censuses.

As energy consumption is highly related to climate rigor, and this is related to degree days [42], cities with heating and cooling degree days higher than 1500 for each regime were selected, resulting in 42 locations, giving priority to those locations with a bigger population, more distance between them and with available weather files.

The selected locations account for 19.05% of the population of Argentina, 2.52% of the population of Brazil, 46.18% of the population of Chile, 13.53% of the population of Mexico and 8.69% of the population of Spain, so the results would cover a significant population from Argentina, Chile and Mexico.

Table 5 shows the selected locations, heating and cooling degree days (HDD and CDD), national climatic zone classification and the minimum thermal transmittance for façades and roofs according to their national requirements for thermal envelopes.

Table 5. Selected locations.

Country	Location	CDD	HDD	Climatic Zone	U Façade	U Roof
Argentina	La Rioja	1368	621	Ia	0.93	0.45
	Santiago del Estero	1311	450			
	Corrientes	1308	192	Ib	1.00	0.45
	Catamarca	1440	531	IIa	0.90	0.45
	Paraná	738	750	IIb	0.99	0.45
	Buenos Aires	732	771	IIIa	1.00	0.48
	Córdoba	705	714			
	Rosario	642	957			
	La Plata	525	1050	IIIb	0.95	0.48
	Mar del Plata	141	1293	IVc	0.85	0.48
Brazil	Curitiba	243	681	1	2.5	2.3
	Ponta Grossa	408	492	2	2.5	2.3
	Santa María	759	462			
	Blumenau	975	138	3	3.7	2.3
	Chapecó	648	387			
	Criciúma	792	252			
	Florianópolis	942	183			
	Porto Alegre	894	372			
Chile	Antofagasta	198	657	1	4	0.84
	Copiacó	453	435	2	3	0.60
	Valparaíso	0	1434			
	Santiago	201	1375	3	1.9	0.47
	Concepción	0	1490	4	1.7	0.38
	Temuco	0	1334	5	1.6	0.33
Mexico	Aguascalientes	589	423	–	0.83	0.83
	Ciudad de Mexico	195	330	–	0.9	0.9
	Guadalajara	717	360	–	0.71	0.71
	Hermosillo	1404	516	–	0.47	0.47
	Juárez	996	1473	–	0.62	0.62
	León	756	189	–	0.71	0.71
	Monterrey	1388	237	–	0.55	0.55
	Puebla	156	429	–	0.83	0.83
	Tijuana	402	657	–	0.71	0.71
Spain	Málaga	864	693	A3	0.94	0.5
	Murcia	1060	849	B3	0.82	0.45
	Palma	873	792			
	Valencia	864	756			
	Alicante	861	771	B4	0.82	0.45
	Córdoba	1080	999			
	Seville	1173	741			
	Barcelona	549	1276	C2	0.73	0.41
	Granada	708	1360	C3	0.73	0.41

2.4.2. Proposals for Walls and Roofs

Five couples of thermal transmittances (façade and roof) were created in order to deal with the requisites of each country. The first couple was established to be the U-values of the case study as the façade and roof materials and configuration are traditionally used in Latin America.

Normative thermal transmittance for the façade and roof for each location were identified Table 5. In Argentina, Brazil, Chile and Spain U-values vary depending on the climatic zone [22,28,36,43]. In Argentina, there is a difference between winter and summer requirements, so the most restricted values were chosen to fulfil both situations. In Mexico, U-values are only defined for certain locations as the normative does not specify a climatic classification [44]; furthermore, there is a change in U-values for buildings of more than three stories high, so the most restricted values were selected.

Finally, the admissible U-values were grouped based on similar ranges, excluding those values corresponding to the case study. Four couples of U-values for the façade and roof were defined in Table 6.

Table 6. Main composition of the five solutions for façade and roof.

Element	Composition	U W/m^{2}K
Wall I	Hollow ceramic brick—15 cm	2.510
Wall II	Double ceramic brick wall—10 cm	1.823
Wall III	Double ceramic brick wall with 1.5 cm of expanded polystyrene 20 kg/m^3	1.101
Wall IV	Hollow ceramic brick-15 cm with fiberglass—3.5 cm	0.821
Wall V	Hollow ceramic brick with 5 cm of expanded polystyrene of 30 kg/m^3	0.503
Roof I	Concrete slab-15cm with alu-zinc roof tiles	1.960
Roof II	Concrete slab-15 cm with fiberglass—3.5 cm	0.854
Roof III	Concrete slab-15 cm with fiberglass—5 cm	0.680
Roof IV	Concrete slab-15 cm with expanded polystyrene of 30 kg/m^3—5 cm	0.526
Roof V	Concrete slab-15 cm with fiberglass—10 cm	0.405

These additional U-values present different materials in the main structure of each element. Excepting Wall II, all additional elements incorporate a thermal insulation layer. Walls are covered on both sides with a plaster layer, the lower side of the roofs are also covered with a plaster layer and the upper side of the slab is a layer of expanded clay.

In order to control the appearance of pathologies, the possibility of condensation risk was analyzed for each wall–roof couple [45]. It was verified that, under normal climatic conditions, there is no superficial or interstitial condensation risk in external walls and roofs.

3. Results

In this chapter, the statistical analysis of the simulation results and the development of the correlational equations are shown. Finally, in order to contrast the results of the equations, they have been compared with the energy consumption resulting from the energy rating schemes of the countries selected.

3.1. Energy Consumption and Linear Regression

The improved energy model was simulated in the 42 locations (scenario R) and also in the 42 locations modifying the U-values of the façade and roof according to national requisites, grouped in five couples (scenario N). The following energy consumption results were obtained, shown in Table 7.

Table 7. Estimated cooling and heating energy consumption ($kWh/m^2/year$).

Country	Location	Cooling_R	Heating_R	Cooling_N	Heating_N	Ave Cool	Ave Heat
Argentina	La Rioja	152.5	150.3	113.6	114.5	81.21 60.87 71.04	238.77 172.57 205.67
	Santiago del Estero	126.8	111.2	94.7	82.9		
	Corrientes	148.6	74	113.5	48.6		
	Catamarca	132.9	120.7	102.5	86.3		
	Paraná	70.3	209	51.3	150.6		
	Buenos Aires	43.6	339.9	32.9	242.5		
	Córdoba	37.8	258.3	27.9	183.6		
	Rosario	64.6	248.4	47.2	181.3		
	La Plata	30.9	364.8	22.6	263.6		
	Mar del Plata	4.1	511.1	2.5	371.8		
Brazil	Curitiba	17.3	148.4	14.3	139.1	54.64 49.64 52.14	110.14 101.40 105.77
	Ponta Grossa	32.5	112.5	28.1	103.3		
	Santa María	62.2	162.7	56.6	149.7		
	Blumenau	53.9	59.2	49.5	53.4		
	Chapecó	14	149.3	11.8	136.7		
	Criciúma	67.6	115	61.7	104.9		
	Florianópolis	92.9	33	85.6	30.4		
	Porto Alegre	96.7	101	89.5	93.7		
Chile	Antofagasta	2.6	65.6	1.7	63.4	3.28 2.22 2.75	363.15 277.76 306.14
	Concepción	0	507.5	0	392.4		
	Copiacó	1.2	238.9	0.6	200.3		
	Santiago	14.8	432.6	10.7	372.4		
	Temuco	0.9	632.4	0.3	506.8		
	Valparaíso	0.2	301.9	0	259.3		
Mexico	Aguascalientes	42.2	36.6	31.5	25	79.93 58.68 69.31	96.87 67.60 82.23
	Ciudad de Mexico	1.8	68	1	44.1		
	Guadalajara	26.9	63.3	17.3	48.6		
	Hermosillo	306.7	19.6	225.7	12.4		
	Juárez	101.7	359.3	72.8	248.3		
	León	36.5	14.7	27.7	7.9		
	Monterrey	194.3	92.2	147.5	56.7		
	Puebla	2.9	71.7	1.4	50.2		
	Tijuana	6.4	146.4	3.2	115.2		
Spain	Málaga	49.9	213.3	40.1	139.6	56.99 43.24 50.12	327.48 220.60 274.04
	Murcia	42.9	303.9	35.1	204.4		
	Palma	71.8	334.8	52.8	245.1		
	Valencia	72	297.2	53.1	206.8		
	Alicante	56.2	231	45.6	153.3		
	Córdoba	60.3	344.4	46.3	233.1		
	Seville	89.4	243.7	63.7	174.1		
	Barcelona	40.4	437.1	28.5	287		
	Granada	30	541.9	24	342		

Average values: first line scenario R, second line scenario N, third line average between both.

It can be observed that, in general, the modification of the original U-values for the walls and roofs in the requisite values of the national legislation, reduces the energy consumption for cooling and heating.

It also can be observed that, although locations within the same climatic classification generally trend to perform in a similar way (i.e., more heating than cooling consumption), there are cases where different trends can be found in the same climatic zone.

A multiple linear regression study was carried out in order to identify the climatic variables more related to heating and cooling energy consumption. Geographic and climatic data for each location were extracted by means of an analysis of the weather file with the Climate Consultant software including the altitude, latitude, monthly average temperature (Tavg), relative humidity (RH), global radiation (RAD), wind speed (Wsp) and sky cover percentage (SKcv).

Furthermore, degree days (DD) were calculated for each location by using the average temperature (Tavg) of each location considering a base temperature of 18 °C [46]. The minimum and maximum design temperature (Td) were also calculated. The maximum design temperature (Td max) was calculated by adding 3.5 °C to the average maximum temperature (Tmax avg) corresponding to the warmest month, and the minimum design temperature (Td min) was calculated by subtracting 4.5 °C from the average minimum temperature (Tmin avg) corresponding to the coldest month according to indications from IRAM 11.603 [45].

Monthly average temperature, relative humidity, global radiation, wind speed and sky cover percentage values were delimited for two different situations in order to determine the set of data more correlated with the variation of energy consumption. The first one considers the annual average; the second considers the average of the three warmest months (average summer—as) to be correlated to cooling consumption and the three coldest months (average winter—aw) to be correlated to heating consumption, shown in Table 8.

Table 8. Input data for the linear regression analysis.

N °	Variable	Definition
1	Cooling	Cooling energy consumption (kWh/m^2)
2	Heating	Heating energy consumption (kWh/m^2)
3	A	Altitude (°)
4	L	Latitude (°)
5	Tavg	Average temperature—monthly (°C)
6	RH	Relative humidity (%)
7	RAD	Global radiation (W/m^2)
8	Wsp	Wind speed (km/h)
9	SKcv	Covered sky (%)
10	CDD	Cooling degree—days
11	HDD	Heating degree—days
12	TDmax	Maximum design temperature (°C)
13	TDmin	Minimum design temperature (°C)
14	Tmax-avg	Average maximum temperature–hottest month (°C)
15	Tmin-avg	Average minimum temperature–coldest month (°C)
16	RHas	Average relative humidity for the three hottest months%
17	RHaw	Average relative humidity for the three coldest months%
18	RADas	Average global radiation for the three hottest months (W/m^2)
19	RADaw	Average global radiation for the three coldest months (W/m^2)
20	Wspas	Average wind speed for the three hottest months (km/h)
21	Wspaw	Average wind speed for the three coldest months
22	SKcvas	Average of covered sky for the three hottest months (%)
23	SKcvaw	Average of covered sky for the three coldest months (%)
24	CDDs	Average cooling degree-days for the three hottest months
25	HDDw	Average heating degree-days for the three coldest months
26	TDmax-s	Maximum design temperature for the three hottest months (°C)
27	TDmin-w	Minimum design temperature for the three coldest months (°C)

The linear regression analysis was performed by SPSS software (version 15.0). The interpretation of the results was based on the consideration of the values of the determination coefficient. R^2 presents null correlation between their variables if the value is 0; very low correlation if the value is between 0.01 and 0.19; low correlation if the value is between 0.2 and 0.39; moderated correlation if the value is between 0.4 and 0.69; high correlation if the value is between 0.70 and 0.89; very high correlation if the value is between 0.9 and 0.99; and a perfect correlation if the value is 1 [47]. Positive values indicate that dependent the variable increases as the independent variable increases; negative values indicate that the dependent variable decreases as the independent variable increases.

The adjusted determination coefficient ($\overline{R}^2$) was used to assess the reliability of the mathematical model. The relevance of the statistical model and the significance of the variables were verified for each case by means of the null p-value hypothesis test in the ANOVA and the probability values. A value ≤ 0.05 was adopted to test the hypothesis that the analyzed variable is of significance.

The method of multiple linear regression analysis was to add the variables forward, in which the software identifies the variables most correlated with the dependent variable (energy consumption), becoming the main variable. Lately, more variables are progressively added in order to increase the R^2, provided that they are influencers of the dependent variable and improve the statistical model, becoming secondary variables. The latest model corresponds to the highest R^2 between the dependent variable and the main and secondary variables, becoming the optimal statistical model. The independent variables not included are neither significant nor influential on the dependent variable.

Results from the multiple linear regression analysis are shown in Tables 9 and 10.

Table 9. Summary of statistical models considering a principal variable + secondary variables.

Condition	Model	Variables	R	R^2	R^2 Corrected	Typical Error
	1	CDD	0.936	0.877	0.874	16.39443
Summer	2	CDD. L	0.951	0.904	0.899	14.67044
	3	CDD. L. CDDs	0.959	0.920	0.913	13.58502
	1	HDD	0.929	0.863	0.860	59.35259
Winter	2	HDD. RADaw	0.946	0.895	0.889	52.78454
	3	HDD. RADaw. Tmin-avg	0.962	0.925	0.920	44.95103
	4	HDD. RADaw. Tmin-avg. Wspaw	0.968	0.938	0.931	41.58207

Table 10. Coefficients for the statistical models considering a main variable + secondary variables.

Condition	Model	Variables	B	Typical Error	Beta	t	Sig
	1	(Constant)	−17.832	4.553	–	−3.917	0.000
		CDD	0.085	0.005	0.936	16.890	0.000
	2	(Constant)	−22.240	4.286	–	−5.189	0.000
Summer		CDD	0.089	0.005	0.987	19.013	0.000
		L	−0.253	0.076	−0.172	−3.310	0.002
	3	(Constant)	−13.616	5.069	–	−2.686	0.011
		CDD	0.123	0.013	1.353	9.513	0.000
		L	−0.221	0.072	−0.150	−3.078	0.004
		CDDs	−0.207	0.076	−0.394	−2.735	0.009
	1	(Constant)	8.300	16.202	–	0.512	0.611
		HDD	0.282	0.018	0.929	15.887	0.000
	2	(Constant)	192.751	56.100	–	3.436	0.001
		HDD	0.242	0.020	0.799	12.359	0.000
		RADaw	−537.190	157.903	−0.220	−3.402	0.002
Winter	3	(Constant)	464.670	83.480	–	5.566	0.000
		HDD	0.162	0.026	0.533	6.163	0.000
		RADaw	−933.288	167.410	−0.382	−5.575	0.000
		Tmin-avg	−14.385	3.622	−0.277	−3.972	0.000
	4	(Constant)	394.912	81.366	–	4.854	0.000
		HDD	0.164	0.024	0.541	6.759	0.000
		RADaw	−871.390	156.525	−0.357	−5.567	0.000
		Tmin-avg	−12.793	3.401	−0.246	−3.762	0.001
		Wspaw	3.233	1.188	0.116	2.722	0.010

It can be observed that degree days are the main variable for heating and cooling energy consumption. The determination coefficients for both situations are 0.86 for heating energy consumption (winter) and 0.87 for cooling energy consumption (summer), indicating a strong variation of the energy consumption with degree days. The inclusion of secondary variables increased the correlation between energy consumption and climatic variables.

The final statistical model presents R^2 = 0.92 for summer and R^2 = 0.93 for winter (heating energy consumption). These results show that cooling energy consumption depends almost completely (92%) on the variation of cooling degree days, latitude and cooling degree days for the three warmest months. Heating energy consumption depends almost completely (93%) on the variation of heating degree days, global radiation of the three coldest months, average minimum temperature and average wind speed for the three coldest months.

Although heating and cooling degree days are the most influential variables on heating and cooling energy consumption, it is observed that not all of the five selected countries use degree days to define climate rigor, shown in Table 11, this is the case for Argentina and Mexico, which use the maximum design temperature and the average maximum temperature, respectively.

Table 11. List of climatic variables used in the rating systems studied.

Country	System	Condition	Main Climatic Variable	Secondary Climatic Variable
ARG	IRAM 11659	Summer	Max design temperature	Solar radiation
	IRAM 11604	Winter	degree days	–
BRA	PBE Edifica	Summer	degree hour	–
		Winter	degree hour	–
CHI	–	Summer	–	–
	CEV	Winter	degree days	Altitude
MEX	NOM-020	Summer	Ave max temperature	Solar radiation
SPA	CEE	Summer	degree days	Solar radiation
		Winter	degree days	Solar radiation

ARG = Argentina, BRA = Brazil, CHI = Chile, MEX = Mexico, SPA = Spain.

In Mexico, furthermore, there is no standard for winter conditions, and in Chile there is no regulation on summer conditions. It is observed by analyzing the cooling energy consumption from the Chilean locations, shown in Table 7, that the climate in these cities has been demonstrated to be slightly rigorous during summer, so there is no need for a summer condition standard. However, heating energy consumption in Mexico was shown to be significant so it would be convenient to include any kind of regulation for winter conditions.

3.2. Equations for Energy Consumption Estimation

Once the statistical models were improved, the equations for the energy consumption estimation of multi-family residential buildings in temperate climates were defined (Equations (1) and (2)):

$$HEC = (0.123 * CDD) - (0.221*L) - (0.207*CDD_s) - 13.616 \tag{1}$$

where HEC is the heating energy consumption, CDD is the cooling degree days (base 18 °C), L is latitude (m) and CDD_s is the average cooling degree days for the three warmest months (base 18 °C).

$$CEC = 394.912 + (0.164 * HDD) - (871.390*RAD_{aw}) - (12.793*T_{amin}) + (3.233*W_{spaw}) \tag{2}$$

where CEC is the cooling energy consumption, HDD is the heating degree days (base 18 °C), RAD_aw is the average global radiation for the three coldest months (W/m^2), T_amin is the average minimum temperature (°C) and W_spaw is the average wind speed for the three coldest months (km/h).

Once the equations are defined, the correlation between the results given by the equations and those obtained by the simulation, shown in Table 7, were compared in order to visualization of their adjustment, shown in Figure 13.

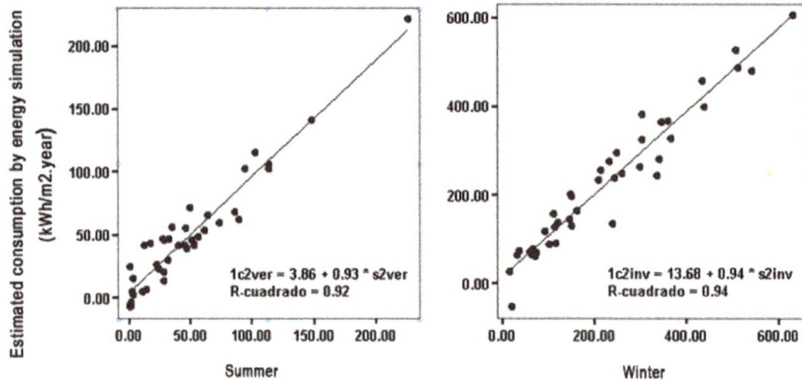

Figure 13. Comparison between energy simulation and statistical model results (kWh/m^2.year).

It can be observed that the R^2 values are 0.92 for cooling and 0.94 for heating, indicating very high reliability in the estimation of energy consumption by the equations.

3.3. Contrasting Predictions from Equations and Established Methods

In order to identify the main similarities and differences between the equations and the existing methods to validate and optimize the model, a comparison between the equations and the Brazilian, Chilean and Spanish existing methods was carried out. The Argentinian and the Mexican method were excluded, as their procedures do not assess thermal loads.

Energy consumption results from equations were compared to the values obtained from the simplified calculation template provided by the PBE Edifica from Brazil and by the CEV from Chile. In the Spanish case, there are some tools, simplified and general methods. For this analysis, the simplified tool CE3X was used.

In the Spanish and Chilean cases, the methods enabled the energy demand to be obtained, so Equation (3) was applied in order to find the energy consumption. The seasonal average performance of the systems has been defined as 1.02 for summer and 1.45 for winter, according to the Institute for the Diversification and Saving of Energy (IDAE), from the Government of Spain [48]:

$$C = D/\eta \tag{3}$$

where C is the final energy consumption (kWh/m^2 year), D is the energy demand (kWh/m^2 year) and η is the seasonal average performance (%)

It was observed that the estimated heating energy consumption by the equation has a difference of 11.2% from CEV results, 21% from CE3x and 210% from PBE Edifica. The estimated cooling energy consumption by the equation has a difference of 43% and 596% from CE3X and PBE Edifica, respectively, shown in Table 12.

Table 12. Comparison between equation results and national calculation tools: PBE for Brazil, CEV for Chile and CE3x for Spain (kWh/m^2 year).

Country	Location	Heating Consumption			Cooling Consumption		
		Equation	Nat. Tool	Δ (%)	Equation	Nat. Tool	Δ (%)
	Blumenau	12.0	9.1	31.9	68.6	8.3	726.4
	Chapecó	43.6	9.1	379.4	37.9	8.3	357.1
	Criciúma	22.7	9.1	149.1	63.2	8.3	661.7
Argentina	Curitiba	77.1	21.5	258.8	2.2	2.6	−17.0
	Florianópolis	6.3	9.1	−30.0	62.4	8.3	651.0
	Ponta Grossa	38.5	10.5	266.9	24.4	4.3	467.4
	Porto Alegre	18.39	9.1	102.0	59.08	8.3	611.8
	Santa Maria	59.53	10.5	466.9	59.26	4.3	1278.2
	Antofagasta	15	43.0	−34.8	–	–	–
	Concepción	227.8	232.0	1.8	–	–	–
Chile	Copiapó	49.3	56.0	12.0	–	–	–
	Santiago	203.5	209.0	2.6	–	–	–
	Temuco	262.8	289.0	9.0	–	–	–
	Valparaíso	155.6	147.5	5.9	–	–	–
	Alicante	124.3	167.5	−25.8	47.1	55.1	−14.5
	Barcelona	178.3	258.75	−31.1	36.5	26.0	40.4
	Córdoba	172.0	167.5	2.7	71.5	46.2	54.8
	Granada	211.5	255	−17.1	43.9	66	−33.5
Spain	Málaga	114.8	123.75	−7.2	52.3	66	−20.8
	Murcia	152.9	167.5	−8.7	51.1	66	−22.6
	Palma	98.3	152	−35.3	57.8	33.1	75.2
	Seville	99.4	167.5	−40.7	102.6	46.2	122.1
	Valencia	113.6	168.75	−32.7	61	66	−7.2

The results indicate that the proposed method is closer to heating energy consumption calculated by the Chilean and Spanish tools and is farther from the cooling energy consumption calculated by the Spanish and Brazilian tools. When comparing the estimated energy consumption results from the equations (Table 7) with the degree days required for the 42 locations (Table 5), it is observed that although there is a very high correlation, for the Brazilian and Spanish cases, the energy consumption calculated by their national tools presents low or very low correlation with degree days, indicating that the proposed method is more consistent with the variation of the local climate rigor (Figure 14).

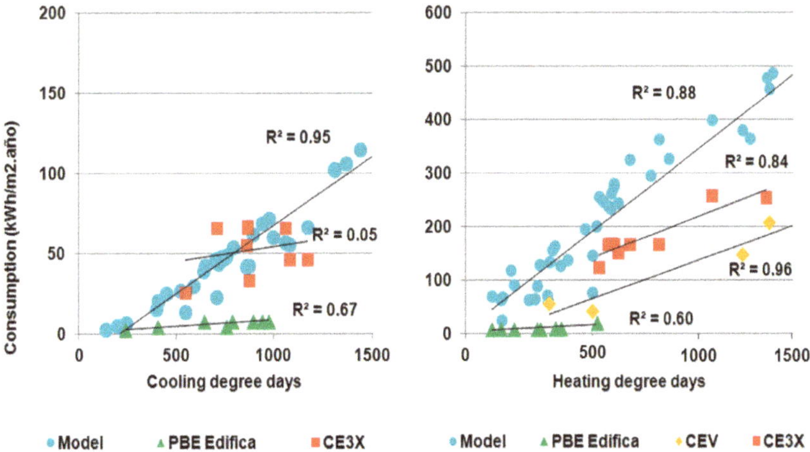

Figure 14. Consumptions estimated by the model and systems of Brazil, Chile and Spain in relation to the degree days of the 42 localities analyzed.

4. Discussion

It is observed that in the Brazilian and Spanish systems, climatic conditions are introduced by means of a selection of the climatic zone (statistically) providing the same energy consumption for different locations within the same climatic zone (i.e., Blumenau, Chapecó, Criciúma, Florianópolis and Porto Alegre in Brazil); the input data in the proposed method are specific climatic data of the location, so it provides further approximation to real conditions. In this regard, the minimum difference to the calculated energy consumption by the Chilean system demonstrates a very high correlation between energy consumption and degree days for each selected Chilean city.

The geometry, building fabric, HVAC systems and user profile characteristics required by the energy rating schemes of the studied countries are similar to the input data of the energy model. However, in the national tools these are input data, while in the proposed method this information is implicit to the equations, so is not possible to modify the building characteristics.

It has to also be highlighted that while this method was built from a user profile developed from a real profile, in the Spanish tool the occupancy and usage profile is predefined for residential buildings with an occupancy thermal load much higher than that of the real profile obtained by surveys. The difference between these profiles is, without doubt, one of the main sources of difference between energy consumption results from the proposed method and the national tools.

The developed methodology is very close to the methods used by Petersdorff, Boermans and Harnisch [14], and Nath and Lamberts [17]. In both cases, an energy model was simulated under different climatic conditions in order to obtain energy consumption data for a statistical analysis. However, the proposed model works with 42 locations when the revised studies worked with three and 18 locations, respectively.

Furthermore, the thermal envelope characteristics in the proposed model is centered in the modification of insulation thickness but Petersdorff, Boermans and Harnisch [14] also vary building typologies, and Nath and Lamberts [17] modified 15 building characteristics without modifying insulation levels. HVAC characteristics are of higher relevance in the model of Nath and Lamberts [17], as it is centered on office buildings.

5. Conclusions

This research describes the methodology developed to create a simplified method to estimate the energy consumption of multi-family housing buildings located in temperate climates, whose input data are just a few climatic variables.

The estimated energy consumptions from the equations present very high values of the determination coefficient ($R^2 = 0.92$ and $R^2 = 0.94$), demonstrating the viability of the application of this method as a tool in quantifying energy consumption.

The proposed method is mainly manual, being easier to use than energy simulation software. It can be implemented in other tools as a spread sheet, in order for even easier use. It can be used by professionals during the building design or reconstruction stages helping to make decisions as it predicts the energy consumption for both heating and cooling, by research centered on the energy performance of buildings, allowing an assessment and qualification of the buildings, and by governmental actors to transfer information to populations in a simple manner, generating awareness about energy efficiency in buildings.

In contrast to other schemes that included climatic conditions based on climatic zones, the proposed method requires some specific local climatic variables as input data, so the energy consumption results are more consistent with the local climatic variations. There is a certain difference in the results from those calculated by national schemes based on climatic zones as they output the same energy consumption for different locations within the same climatic zone. In contrast, results from the proposed method and the Chilean one, as they vary with degree days, are very close.

The multiple linear regression analysis demonstrates that, in general, degree days is the most influential variable on energy consumption in residential buildings. In summer conditions, latitude and average degree days for the three warmest months are also influential. In winter conditions, the average global solar radiation for the three coldest months, average minimum temperature and average wind speed for the coldest months have to be taken into account.

Author Contributions: G.R.-N., P.M.-M. and J.D.C. conceived and designed the experiments. G.R.-N. performed the experiments. G.R.-N., P.M.-M. and J.D.C. analyzed the data. G.R.-N. and P.M.-M. wrote the paper.

Funding: The National University of La Plata through the project "Sustainable Buildings Rating for adaptation and mitigation of Climate Change" (11/U141) funded this research. The main author received the financial support of the Doctoral Scholarship for strategic areas of the National Council of Scientific and Technical Research (CONICET) from Argentina. Furthermore, the main author received the financial support of the Scholarship Programme for Training Abroad (BecAR) of the Ministry of Modernization of Argentina and a Scholarship for International mobility of doctoral students for the development of co-supervised doctoral thesis from the III Proper Teaching Plan of the University of Seville (Action 2.2.3 2018), which made able the stay in Seville. The wording of this paper is part of the projects "Rehabilitación Ecoeficiente de Edificios y Barriadas" Ref: (CCPI 2015/006) and Ref. 3557/0632 of the Universidad de Sevilla, Spain, financed by GRUPO PUMA S.L.

Acknowledgments: The authors would like to thank the National University of La Plata (UNLP) for the license of the SPSS software and the University of Seville for the license of the Design Builder software and the support for co-supervision of doctoral thesis of the III Proper Teaching Plan of the University of Seville.

Conflicts of Interest: The authors declare no conflict of interest. The funders had no role in the design of the study; in the collection, analyses, or interpretation of data; in the writing of the manuscript, or in the decision to publish the results.

References

1. Pérez-Lombard, L.; Ortiz, J.; RGonzález, R.; Maes, I.R. A review of benchmarking, rating and labelling concepts within the framework of building energy certification schemes. *Energy Build.* **2009**, *41*, 272–278. [CrossRef]
2. Walsh, A.; Cóstola, D.; Labaki, C.L. Review of methods for climatic zoning for building energy efficiency programs. *Build. Environ.* **2017**, *112*, 337–350. [CrossRef]
3. Baldwin, R.; Leach, S.J.; Doggart, J.; Attenboroug, M. *BREEAM 1/90: An Environmental Assessment for New Office Designs*; IHS BRE: Garston, UK, 1990.
4. Feist, W. The Passive House in Darmstadt–Kranichstein during Spring, Summer, Autumn and Winter. 2006. Available online: https://passiv.de/former_conferences/Kran/Passive_House_Spring_Winter.htm (accessed on 17 November 2017).
5. Cole, R.J.; Rousseau, D.; Theaker, I.T. *Building Environmental Performance Assessment Criteria: Version 1 Office Buildings*; UBC School of Architecture, Environmental Research Group: Vancouver, BC, Canada, 1993.
6. U.S. Energy Information Administration. *Energy Statistics*; U.S. Energy Information Administration: Washington, DC, USA, 2017. Available online: https://www.eia.gov/beta/international/ (accessed on 5 May 2018).
7. Peel, M.C.; Finlayson, B.L.; McMahon, T.A. Updated world map of the Koppen-Geiger climate classification. *Hydrol. Earth Syst. Sci.* **2007**, *11*, 1633–1644. [CrossRef]
8. European Council. *Directive 2002/91/EC of the European Parliament and of the Council of 16 December 2002 on the Energy Performance of Buildings*; European Council: Brussels, Belgium, 2002.
9. Mercader-Moyano, P.; Claro-Ponce, J. Sistemas de Certificación en Clima Templado. *ARQUISUR Rev.* **2017**, *7*, 62–77. [CrossRef]
10. Netto, G.R. Sistemas de calificación edilicia en Latinoamérica. Master's Thesis, Universidad Nacional de La Plata, La Plata, Argentina, 2017.
11. Krsti, H.; Teni, M. Review of Methods for Buildings Energy Performance Modelling. *IOP Conf. Ser. Mater. Sci. Eng.* **2017**, *245*, 042049. [CrossRef]
12. Crawley, D.B. Building Performance Simulation: A Tool for Policymaking. Ph.D. Thesis, The University of Strathclyde, Glasgow, UK, 2008.
13. Melo, A.P.; Cóstola, D.; Lamberts, R.; Hensen, J.L.M. Assessing the accuracy of a simplified building energy simulation model using BESTEST: The case study of Brazilian regulation. *Energy Build.* **2012**, *45*, 219–228. [CrossRef]

14. Petersdorff, C.; Boermans, T.; Harnisch, J. Mitigation of CO2 emissions from the EU-15 building stock: Beyond the EU Directive on the Energy Performance of Buildings. *Environ. Sci. Pollut. Res.* **2006**, *13*, 350–358. [CrossRef]

15. Wong, S.L.; Wan, K.K.; Lam, T.N. Artificial neural networks for energy analysis of office buildings with daylighting. *Appl. Energy* **2010**, *87*, 551–557. [CrossRef]

16. Korolija, I.; Zhang, Y.; Marjanovic-Halburd, L.; Hanby, V.I. Regression models for predicting UK office building energy consumption from heating and cooling demands. *Energy Build.* **2013**, *59*, 214–227. [CrossRef]

17. Lopes, M.; Lamberts, R. Development of a metamodel to predict cooling energy consumption of HVAC systems in office different climates. *Sustainability* **2018**, *10*, 4718. [CrossRef]

18. Mercader-Moyano, P.; Santiago, M.O.; Ramírez-de-Arellano, A. Modelo de cuantificación del consumo energético en edificación. *Mater. de Construcción* **2012**, *62*, 567–582. [CrossRef]

19. Sánchez-García, D.; Rubio-Bellido, C.; Pulido-Arcas, J.; Guevara-García, F.; Canivell, J. Adaptive Comfort Models Applied to Existing Dwellings in Mediterranean Climate Considering Global Warming. *Sustainability* **2018**, *10*, 3507. [CrossRef]

20. Ministerio de la Vivienda. Gobierno de España. *Real Decreto 314/2006, de 17 de marzo, por el que se aprueba el Código Técnico de la Edificación*; Ministerio de la Vivienda. Gobierno de España: Madrid, Spain, 2006.

21. Escandón, R.; Suárez, R.; Sendra, J.J. On the assessment of the energy performance and environmental behaviour of social housing stock for the adjustment between simulated and measured data: The case of mild winters in the Mediterranean climate of southern Europe. *Energy Build.* **2017**, *152*, 418–433. [CrossRef]

22. Ministerio de la Vivienda. Gobierno de España. *Documento Básico HE: Ahorro de Energía*; Ministerio de la Vivienda. Gobierno de España: Madrid, Spain, 2017.

23. Ministerio de Infraestructura. Gobierno de Buenos Aires. *Decreto n° 1.030–Reglamento de aplicación de la Ley n° 13.059*; Ministerio de la Vivienda. Gobierno de España: Madrid, Spain, 2010.

24. Programa Brasileiro de Etiquetagem. Available online: http://www.pbeedifica.com.br/ (accessed on 23 February 2009).

25. Ministerio de Vivienda y Urbanismo. Gobierno de Chile. *Manual de Procedimientos Calificación Energética de Viviendas en Chile*; Ministerio de Vivienda y Urbanismo. Gobierno de Chile: Santiago, Chile, 2018.

26. Sociedad Hipotecaria Federal. Gobierno de México. Available online: https://www.gob.mx/shf/documentos/ecocasa (accessed on 6 October 2012).

27. Ministério do Desenvolvimento, Indústria e Comércio Exterior do Brazil. *Regulamento Técnico da Qualidade para o Nível de Eficiência Energética de Edificações Residenciais*; Ministério do Desenvolvimento, Indústria e Comércio Exterior do Brazil: Rio de Janeiro, Brazil, 2012.

28. Ministerio de Vivienda y Urbanismo. Gobierno de Chile. *Reglamentación Térmica. Ordenanza General de Urbanismo y Construcciones*; Diario Oficial; Ministerio de Vivienda y Urbanismo. Gobierno de Chile: Santiago, Chile, 2005.

29. Foro de Ministros y Autoridades Máximas de la Vivienda y el Urbanismo de América Latina y el Caribe (MINURVI). *América Latina y el Caribe: Desafíos, dilemas y compromisos de una agenda urbana común*; Naciones Unidas: Santiago, Chile, 2016.

30. Economic Commission for Latin America and the Caribbean (ECLAC). ECLAC United Nations. Available online: https://www.cepal.org/en (accessed on 9 March 2018).

31. Economic Comision for Latin American and the Caribbean (ECLAC). CEPALSTAT|Databases and Statistical Publications. Economic Comision for Latin American and the Caribbean. Available online: http://interwp.cepal.org/cepalstat (accessed on 9 March 2018).

32. Ministerio de Fomento. Reino de España. *Análisis de las características de la edificación residencial en España 2011*; Ministerio de Fomento: Madrid, Spain, 2014.

33. INDEC. Instituto Nacional de Estadística y Censos. República Argentina. Available online: https://www.indec.gob.ar/ (accessed on 9 March 2018).

34. Instituto Brasileiro de Geografia e Estadistica. Pesquisa Nacional por Amostra de Domicílios Contínua—PNAD Contínua. Available online: https://www.ibge.gov.br/estatisticas/sociais/habitacao (accessed on 9 March 2018).

35. Comisión Nacional de Vivienda. Mexico. *Resumen del Estudio de Mercado de Vivienda Existente*; Deutsche Gesellschaft für Internationale Zusammenarbeit (GIZ): Mexico, Spain, 2012.

36. Associação Brasileira de Normas Técnicas. *NBR 15.220—desempenho térmico de edificações*; Associação Brasileira de Normas Técnicas: Rio de Janeiro, Brazil, 2008.
37. Instituto Argentino de Normalización y Certificación. *IRAM 11.604—Aislamiento térmico de edificios. Ahorro de energía en calefacción. Coeficiente volumétrico G de pérdidas de calor*; Instituto Argentino de Normalización y Certificación: Buenos Aires, Argentina, 2004.
38. Roriz, M.; Ghisi y, E.; Lamberts, R. Bioclimatic zoning of Brazil: A proposal based on the Givoni and Mahoney methods. In Proceedings of the 16th International Conference on Passive and Low Energy Architecture, Brisbane, Australia, 1999; Available online: http://www.infohab.org.br/acervos/advanced-search (accessed on 12 June 2015).
39. DesignBuilder Software Ltd. DesignBuilder. Available online: https://designbuilder.co.uk/index.php (accessed on 3 September 2015).
40. EnergyPlus. Weather Data|Energy Plus. Available online: https://energyplus.net/weather (accessed on 3 September 2015).
41. American Society of Heating, Refrigerating and Air-Conditioning Engineers (ASHRAE). *ANSI/ASHRAE ASHRAE Guideline 14-2014 Measurement of Energy and Demand Savings*; ASHRAE: Atlanta, GA, USA, 2014.
42. Lucas, I.B.; Hoesé, L.; Facchini, M.L. *Análisis de la relación entre clima y consumo energético residencial en la ciudad de San Juan*; Energías renovables y medio ambiente: Buenos Aires, Argentina, 2011; Volume 28, pp. 17–25.
43. Instituto Argentino de Normalización y Certificación. *IRAM 11.605—Acondicionamiento térmico de edificios: Condiciones de habitabilidad en edificios: Valores máximos de transmitancia térmica en cerramientos opacos*; Instituto Argentino de Normalización y Certificación: Buenos Aires, Argentina, 2001.
44. Secretaría de Energía. Norma Oficial Mexicana. *NOM 020 ENER 2011—Eficiencia Energética en Edificaciones. Envolvente de Edificios para Uso Habitacional*; Secretaría de Energía. Norma Oficial Mexicana: Mexico City, Mexico, 2011.
45. Instituto Argentino de Normalización y Certificación. *IRAM 11.625—Aislamiento térmico de edificios: Verificación del riesgo de condensación de vapor de agua superficial e intersticial en los paños centrales de muros exteriores, pisos y techos de edificios en general*; Instituto Argentino de Normalización y Certificación: Buenos Aires, Argentina, 1996.
46. Instituto Argentino de Normalización y Certificación. *IRAM 11.603—Acondicionamiento térmico de edificios: Clasificación bioambiental de la Republica Argentina*; Instituto Argentino de Normalización y Certificación: Buenos Aires, Argentina, 2012.
47. Pearson, K. Notes on the history of correlation. *Biometrika* **1920**, *13*, 25–45. [CrossRef]
48. Instituto para la Diversificación y Ahorro de la Energía. *Escala de calificación energética para edificios de nueva construcción*; IDEA: Madrid, Spain, 2009.

Article

Window Design of Naturally Ventilated Offices in the Mediterranean Climate in Terms of CO$_2$ and Thermal Comfort Performance

Hardi K. Abdullah * and Halil Z. Alibaba *

Department of Architecture, Faculty of Architecture, Eastern Mediterranean University, via Mersin 10, Famagusta 99628, North Cyprus, Turkey
* Correspondence: hardi.abdullah@su.edu.krd (H.K.A.); halil.alibaba@emu.edu.tr (H.Z.A.);
  Tel.: +90-533-821-4645 (H.K.A.); +90-533-863-0881 (H.Z.A.)

Received: 29 November 2019; Accepted: 6 January 2020; Published: 8 January 2020

Abstract: Natural ventilation through window openings is an inexpensive and effective solution to bring fresh air into internal spaces and improve indoor environmental conditions. This study attempts to address the "indoor air quality–thermal comfort" dilemma of naturally ventilated office buildings in the Mediterranean climate through the effective use of early window design. An experimental method of computational modelling and simulation was applied. The assessments of indoor carbon dioxide (CO$_2$) concentration and adaptive thermal comfort were performed using the British/European standard BS EN 15251:2007. The results indicate that when windows were opened, the first-floor zones were subjected to the highest CO$_2$ levels, especially the north-facing window in the winter and the south-facing window in the summer. For a fully glazed wall, a 10% window opening could provide all the office hours inside category I of CO$_2$ concentration. Such an achievement requires full and quarter window openings in the cases of 10% and 25% window-to-floor ratios (WFR), respectively. The findings of the European adaptive comfort showed that less than 50% of office hours appeared in category III with cross-ventilation. The concluding remarks and recommendations are presented.

Keywords: window design; natural ventilation; indoor air quality; carbon dioxide (CO$_2$) concentration; thermal comfort; adaptive comfort model; office building; the Mediterranean climate

1. Introduction

In urban areas, people spend most of their time (nearly 90%) indoors while performing different daily activities, where the concentration of most indoor pollutants is about 20% higher than in the outdoor environment [1]. Therefore, maintaining comfortable and healthy conditions for occupants is one of the major building tasks. Indoor air quality (IAQ) has a significant impact on human health and comfort. Modern lifestyle requires paying more attention to the provision of better thermal comfort and healthier indoor conditions for occupants, while advancements in technology and mechanical systems have created the means of achieving this goal. However, sustainability standards and green building guidelines require less dependence on active strategies to minimise energy consumption, and consequently, reduce buildings' carbon footprints.

Carbon dioxide (CO$_2$) is one of the most common gases found in our atmosphere. It can be used as a good indicator of human bio-effluent concentration. An indoor CO$_2$ measurement provides a dynamic measure of the balance between carbon dioxide generation in the space, representing occupancy, and the amount of low CO$_2$ concentration in the outside air introduced for ventilation. Air movement has a significant influence on perceived indoor air quality [2]. Researchers claim that the air tightening within an occupied zone of air-conditioned spaces will result in complaints of unsatisfactory indoor air. Field studies suggest that the elevated airspeed within an occupied zone can possibly achieve thermal comfort even at higher temperatures and improve the perceived indoor air quality [3].

In recent studies, the utilization of natural ventilation, as a prevalent and effective passive strategy, to remove indoor pollutants and maintain indoor air quality along with indoor thermal comfort of various building programs is being challenged. The findings of previous studies recommend conflicting objectives and emphasise the need to pursue a more integrative approach to indoor environmental quality by tackling more than one criteria simultaneously [2].

Windows are the main and most popular means in which natural ventilation can be allowed into a building's indoor spaces. Natural ventilation through windows can be based on pressure difference (also called wind-driven natural ventilation) or thermal difference (in single-sided ventilation or when placing windows or openings at different heights in cross-ventilation) between inside and outside or between the openings [4]. Occupant-controlled windows are considered an effective method for maintaining indoor air quality and thermal comfort conditions. Window-based natural ventilation can replace mechanical ventilation and air condition systems (in free-running buildings or periodically) [5], thus reducing a significant amount of energy consumption and CO_2 emissions [6]. Accordingly, window design has a strong relationship with natural ventilation performance in different types of buildings. Window design is an early decision task of architects that requires sufficient knowledge supported by experiments and quantitative data.

Studies confirm that window-based natural ventilation is an inexpensive and practical method to bring fresh air into internal spaces and enhance indoor air quality and thermal comfort [6–9]. Yet, opening windows in the warm months may result in indoor overheating; consequently, an "indoor air quality-thermal comfort" dilemma exists [10–12]. Previous studies have mainly studied natural ventilation performance only in terms of indoor air quality or thermal comfort. This study attempts to address the "indoor air quality-thermal comfort" dilemma of naturally ventilated office buildings in the Mediterranean climate through the effective use of early window design. It examines the potential performance of single-sided and cross-ventilation by investigating different window design scenarios, including window size, orientation, location (different floor levels), and possible opening behaviour (by occupants). Architects unconsciously limit the amount of airflow coming into a building from openings when they choose a particular window size, orientation, and type in the early design stage. Nowadays, for instance, modern office buildings with large glazed walls have limited windows for natural ventilation, or a particular type of windows has a limited opening area, which might reduce ventilation and cooling capabilities of ambient air, especially in naturally ventilated buildings. An adequately designed window can lead to maximising the free-running period—no mechanical systems are used for ventilation and air-conditioning—and thus saving a considerable amount of energy. Therefore, architects need to understand the traces of window design decisions in terms of natural ventilation performance. Accordingly, the outcomes of this research can help architects to make informed choices when they decide on the different parameters related to window design considering both indoor air and thermal conditions, simultaneously, in the early design stage.

2. The Effect of Building Envelope Design on Indoor Environmental Performance

A building envelope separates the indoor spaces from the outdoor environment. It is an external layer of the building that protects the internal environment from harsh environmental conditions and facilitates climate control. Therefore, the climatic design of a building envelope has an impact on its indoor air quality, thermal and visual performance, and energy consumption. In the Mediterranean climate, it is important to limit the amount of heat gain through the design of the building envelope and utilise effective natural ventilation strategies to cool down the internal spaces in the summer months.

Turkish researchers [13] examined the impact of passive solar building components on the energy performance of residential units in Turkey's different climates. The results revealed that the building aspect ratio has less influence on the total energy demand compared to the window size and insulation materials. Moreover, compact forms and large-size windows are the most preferable combination in the cool climates, while the situation is the total reverse in the warm climates. Based on the concept of passive and non-passive spaces developed by Baker and Steemers [14] and adopted by Steadman et al. [15] for the energy classification of built forms, Ratti et al. define a 'passive zone' as one that can successfully be treated using passive strategies [16]. According to empirical observations, a 'passive zone' is considered twice the ceiling height. A similar study [17] proved that minimising the building's shape coefficient reduces heat loss in winter; however, it negatively affects the 'passive zone' by reducing the availability of natural ventilation and daylight. Thus, an envelope less exposed to the outside environment increases the energy demand for artificial lighting and ventilation. While the 'passive zone' has been considered a better indicator for energy consumption [15], it can consume even more energy compared to the non-passive zone if the glazing is not designed to prevent overheating in the summer and heat loss in the winter.

Moreover, researchers [18] studied various building forms and plan layout designs to access passive strategies in relation to thermal comfort and natural ventilation in a university building. They found that plans longer than 15 m could lower the effect of natural ventilation to provide thermal comfort. Other studies examined the potential of different building forms to reduce solar radiation [19], thermal performance, and energy use [20]. Studies confirmed that room height has a considerable influence on energy demand, such that the energy consumption increases by 1% for each 10 cm increase in ceiling height [21]. Although a reduction in ceiling height offers less exposed surface areas, it can result in higher indoor temperatures and consequently, less thermally comfortable indoor spaces, especially in the warm and hot climates [22]. The building orientation also has a considerable effect on energy consumption and thermal comfort as it is implicated in the levels of solar radiation, daylighting, and air movement [23]. Regardless of building form, buildings arranged longitudinally along the south and north require 10% less energy than those aligned longitudinally along the east and west in a hot and humid climate [20]. A study [24] assessed both IAQ and thermal comfort, as one package, in recently built energy-efficient houses. The findings indicate that in these airtight houses, mechanical ventilation has to be working constantly to maintain indoor environmental conditions. Another study combined objective environmental variables and subjective comfort evaluation to assess indoor air quality and thermal comfort based on Weber/Fechner's law and Predicted Mean Vote (PMV) [25].

Previous studies focused less on examining the relationships between window design and natural ventilation, as well as the effect of different window design parameters on the indoor CO_2 concentration and thermal comfort performance. A larger part of existing research concentrates either on the reduction of energy demand [6,13,26,27] or improving thermal comfort levels by exploring a particular building component [28,29].

Window Design in Relation to CO_2 and Thermal Comfort in Naturally Ventilated Buildings

Windows are designed at the early architectural design phase where designers decide on most of the envelope-related elements. These decisions have a significant influence on building performance in terms of indoor air quality, thermal comfort, visual comfort, daylighting, and eventual energy consumption [6,20,30,31]. Different climatic conditions require specific envelope design considerations to achieve an environmentally responsive envelope design. In the Mediterranean climate, there is a need to limit the amount of solar heat gain in the summer and heat loss in the winter, especially through window openings. Besides, window-based natural ventilation can be exploited efficiently to cool down internal spaces in the warm months.

Natural ventilation in buildings mainly occurs through intended envelope openings (e.g., windows or doors) and infiltration (leakage of the building surfaces) as a result of differences in pressure between the inside and outside [32]. In unconditioned spaces, therefore, natural ventilation is the only method to dilute indoor air contaminants, particularly the carbon dioxide exhaled by occupants. There are several strategies for natural ventilation, such as single-flow ventilation, cross-flow ventilation, internal ventilation, and the thermal chimney effect. This study examines single-side and cross-flow ventilation strategies with different window design strategies. Numerous studies have assessed various window parameters in relation to particular building performance objectives or multiple performance criteria. Most countries follow certain building code and design guidelines to specify the window-to-wall ratio (WWR) or window-to-floor area ratio (WFR). The impact of WWR on different building performance goals has been studied more frequently [33–38].

Alibaba [39] studied the heat and airflow behaviour of naturally ventilated offices in a Mediterranean climate (i.e., Famagusta, North Cyprus). One aspect of the study was examining the effect of different window-to-wall ratios and window openings on the air change rates (ach) per hour. The maximum ach was achieved when the building had a 100% WWR with fully open windows, whereas the minimum ach was recorded in the case of 10% WWR with a 20% window opening. Mora-Pérez et al. [6] studied natural ventilation design decisions concerning energy efficiency and CO_2 emission of a residential building in the Mediterranean region. The authors claimed that the building's natural ventilation behaviour was improved by 9.7% with a new opening alternative.

Research into the indoor air quality of naturally ventilated high-occupancy research student offices at Beijing University, China [40] investigated the carbon dioxide concentration and indoor climate (i.e., dry-bulb air temperature and relative humidity) during the heating period. The quantitative measurements show that the indoor CO_2 level exceeded the threshold of 1000 ppm throughout most of the occupied time each day. The average exposure to CO_2 concentration over the threshold was 3.68 h per occupant per day. Therefore, these offices do not meet the IAQ requirements and users tend to suffer health consequences. Laska and Dudkiewicz [41] studied CO_2 concentration in a naturally ventilated lecture room at the Wroclaw University of Science and Technology, Poland. The city is characterised by a mild and moderately warm climate. The collected data from field measurements validated a model previously derived for school classrooms [42]. The authors argue that this model is also applicable for calculating the CO_2 concentration in auditorium lecture rooms where the occupants are the main source of pollution. The measured values of CO_2 concentration were compared to the acceptable level of carbon dioxide defined in the European Standard 13779:2008 and a questionnaire survey based on personal discomfort. The results of this experimental study indicate that during a 90-min lecture, the concentration was within the permissible levels and the occupants were satisfied. However, when the room was fully occupied, the indoor environment failed to provide suitable health conditions. These conclusions indicate that naturally ventilated indoor spaces need to be regularly aired to maintain the comfort conditions and productivity of users.

In the literature survey, researchers mainly depend on CO_2 concentration (ppm) as a proper indicator to assess natural ventilation performance [12,30,40–43] in reference to the 1000 ppm threshold defined by the World Health Organisation (WHO) [44]. In other words, CO_2 levels higher than 1000 ppm denote insufficient ventilation. Exceeding this threshold can cause sick building syndrome (SBS) problems for residents, such as headaches and respiratory problems [7,45–48]. Nevertheless, in naturally ventilated buildings, where occupants have full access to openable windows, minimal indoor CO_2 levels might be preferable. Considering that CO_2 concentration in the air is about 350–450 ppm, appropriately designed windows and opening portions can reduce the internal CO_2 level.

Researchers in Spain [49] investigated the potential of adaptive thermal comfort for existing dwellings in the Mediterranean climate. The authors declared that both EN 15251:2007 and American Society of Heating, Refrigerating, and Air-Conditioning Engineers (ASHRAE) 55-2017 are applicable to the considered conditions and both standards presented comparable results, noting that EN 15251:2007 standard can predict worse conditions than the American model. Other researchers [50] confirm that regional adaptive comfort indicators showed more reliable results than the ASHRAE adaptive model for school buildings in the Mediterranean climate. A similar study in the same country [51] applied adaptive thermal comfort in Mediterranean office buildings. They found that natural ventilation through window openings (manual or mechanical) provided up to 30% more occupancy hours that are comfortable based on the EN 15251:2007 standard, and with window-material improvements, that percentage could be raised to more than 50%.

Salvalai et al. [52] studied the thermal comfort and energy performance of several low-energy cooling concepts for office buildings in six different European climate zones. A series of dynamic simulations were performed based on the PMV (ISO 7730:2005) and adaptive (EN 15251:2007) thermal comfort models. The findings indicate that natural ventilation has a greater potential for the Northern and Central parts of Europe compared to Southern Europe due to the presence of higher ambient air temperatures in the later climate. Even in European climates, solely implementing passive cooling methods has its limitations in terms of achieving thermal comfort. From an architectural perspective, an adequate knowledge on window design and natural ventilation relationship, considering a particular local condition, can guide architects toward selecting an optimum window design that maximises natural ventilation and passive cooling performance [52,53]. Croitoru et al. [54] investigated the thermal comfort of a low energy office building in the temperate climate of Romania. The study compared real-life experimental results with the subjective responses from a questionnaire on the thermal sensation votes. The thermal comfort results placed the free-running building in Category I and Category II of European adaptive comfort (EN 15251:2007).

3. Materials and Methods

An experimental method of computational modelling and simulation techniques was used to collect and analyse numerical data. This study phase encompasses the selection of building performance simulation (BPS) tool, describing features of the hypothetical building case, and identifying performance criteria and assessment methods. Figure 1 illustrates the research methodology flowchart.

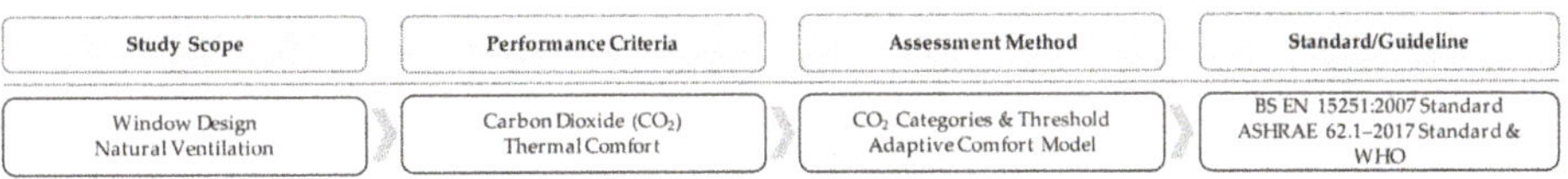

Figure 1. The methodology flowchart.

3.1. Building Performance Simulations (BPS)

Developed by Environmental Design Solutions Limited (EDSL), TAS Engineering software version 9.4.4 [55] is used to conduct the computational thermal simulations and fulfil the aim of the study. TAS Engineering software is a complete solution for the dynamic simulation and thermal analysis of buildings. TAS software is "an industry-leading building modelling and simulation tool capable of performing hourly dynamic thermal simulation for the world's largest and most complex buildings" [55]. As a complete solution for the thermal simulation of new and existing buildings, the software scope facilitates a methodical workflow. The '3D Modeller' can create building models for simulation and performing daylight analyses. The 'Building Simulator' allows for the addition of apertures, internal gains, constructions, and the performance of dynamic simulations. Finally, the 'Result Viewer' is for storing, viewing and exporting hourly results in both 2D and 3D.

3.2. Climate Analysis of Famagusta

This study phase of the methodology identifies the contextual climate conditions through comprehensive weather classification and analysis. Although several methods have been introduced by scholars, the Köppen-Geiger Climate Classification [56,57] is considered one of the most reliable and widely used systems for classifying climates. This system divides climates into five main climate groups, with each group further divided based on the monthly and annual averages of precipitation and temperature patterns. According to the Köppen-Geiger Climate Classification system, the five main climate groups are *A* (tropical), *B* (arid), *C* (warm temperate), *D* (continental), and *E* (polar).

Based on the Köppen-Geiger Climate Classification, Famagusta's climate (latitude 35.0° N and longitude 33.0° E) is the Csa (Mediterranean climate), which is characterised by dry and hot summers and rainy, rather changeable, winters. The warm period starts in May and lasts until the end of

September. While the cool period is between November and March, April and October are rather moderate months.

The driving forces in natural ventilation are temperature and wind; therefore, the significant factors are the outdoor and indoor conditions that should be considered when studying natural ventilation. On average, July is the warmest month in the year, while the hottest temperature occurs in July and August with a mean daily outdoor temperature of 28 °C. January is the coldest month in the year, for which the average daily outdoor temperature is about 11 °C. The average annual day temperature is 25 °C and the average annual night temperature is 13 °C. Temperatures vary significantly between day and night, which ranges between, approximately, 10 °C in the winter to 12 °C in the summer. Furthermore, December and June represent the most and least humid months of the year with approximately 73% and 64% humidity ratios, respectively. The average annual percentage of relative humidity is about 69%. The city's dominating winds are from the west, north in winter and west, south in summer. These wind directions may improve the effectiveness of natural ventilation when the windows are aligned with these orientations. For reference, the windiest and calmest days are recorded in February and September with the daily average wind speed of 5.2 m/s and 3.3 m/s, respectively. Tables 1 and 2 outline the climatic conditions of the study location. Figure 2 shows the wind rose of Famagusta.

Table 1. Monthly average temperatures and relative humidity based on Famagusta weather data.

Month	Average Temp (°C)	Mean Max Temp (°C)	Mean Min Temp (°C)	Temperature Difference (°C)	Relative Humidity (%)
January	10.9	16.4	6.9	9.7	72.8
February	12.8	16.4	6.5	10.3	71.7
March	14.0	18.4	7.8	10.8	72.8
April	16.2	22.2	10.5	11.8	70.7
May	21.4	26.5	14.2	12.2	67.3
June	26.0	30.6	18.4	13.2	64.3
July	28.4	33.1	21.1	12.1	65.0
August	28.4	33.3	21.4	12.1	67.3
September	25.7	31.1	16.4	13.1	66.6
October	22.8	27.2	15.3	11.8	67.5
November	17.9	22.0	11.0	10.8	70.0
December	13.7	17.6	7.5	9.5	73.2

Table 2. Monthly average wind speed and predominant wind directions based on Famagusta weather data.

Month	Average Wind Speed (m/s)	Percentages of Predominant Wind Directions (%)			
		North	East	South	West
January	5.0	35	25	10	30
February	5.1	30	20	13	37
March	4.6	30	13	12	45
April	4.0	22	13	15	50
May	3.5	18	7	15	60
June	3.4	10	5	20	65
July	3.5	10	3	27	60
August	3.4	10	3	25	62
September	3.3	15	5	15	65
October	3.6	35	10	10	45
November	4.3	45	20	10	25
December	4.8	38	20	12	30

Note: The colour scheme indicates first (darker) and second (lighter) predominant wind directions.

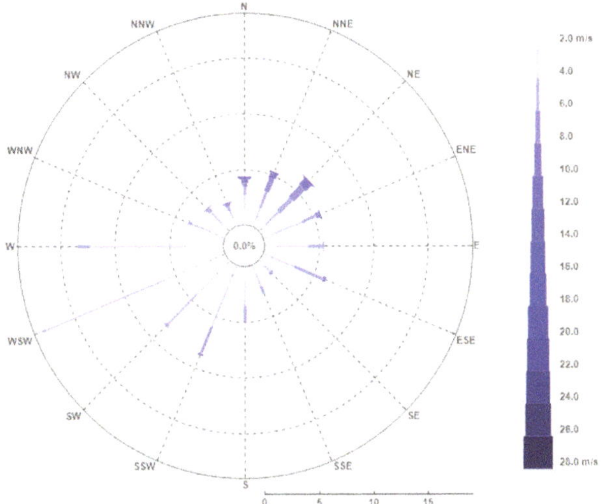

Figure 2. The wind rose of Famagusta based on its weather file data [55].

3.3. Building Case and Window Design Features

This study targets the early envelope, particularly window, design of office buildings in the Mediterranean climate. To replicate common building designs in the study location and to test different window orientations and floor locations, a hypothetical building was designed as a three-storey office building with four thermal zones on each floor, as presented in Figure 3. Each zone had an area of 50.0 m² with a 1:1 length-to-width ratio, also called the space aspect ratio (7.1 m × 7.1 m). The height of the ceiling was fixed at 3.0 m as the normal ceiling height recommended by the local building design regulation of the study location [58]. The minimum window-to-floor ratio accepted by the North Cyprus Chamber of Architects is 10% WFR [58]. The other scenarios included 25% and 50% (full glass in this building case). The natural ventilation patterns were single-side ventilation in the cases of 10% and 25% WFR, as well as cross ventilation in the case of 50% WFR. The authors tested various aperture opening scenarios ranging from closed to fully opened windows for the different orientations. It is important to mention that neither external solar shadings nor internal blinds were used to reflect common office design practice or the worst status of windows in response to excessive solar impact. Table 3 presents the considered building and window design parameters as well as different simulation scenarios. Tables 4 and 5 show the transparent and opaque construction materials and their specifications that are commonly utilised for building construction in North Cyprus.

Table 3. Building geometric parameters and various simulation scenarios.

Building Geometric Parameters	Unit	Simulation Scenarios
Space aspect ratio (L/W)	–	1:1
Space clear height	(m)	3.0
Floor location	–	Ground, first, and second floor
Window-to-floor ratio (WFR)	(%)	10, 25, 50 (fully glazed wall)
Window orientation	–	North, east, west, and south
Window opening ratio	(%)	0 (closed), 10, 25, 50, 75, 100 (fully open)
Window shading ratio	(%)	N/A
Natural ventilation strategy	–	Single-side for 10% & 25% WFR Cross-flow for 10% & 50% WFR

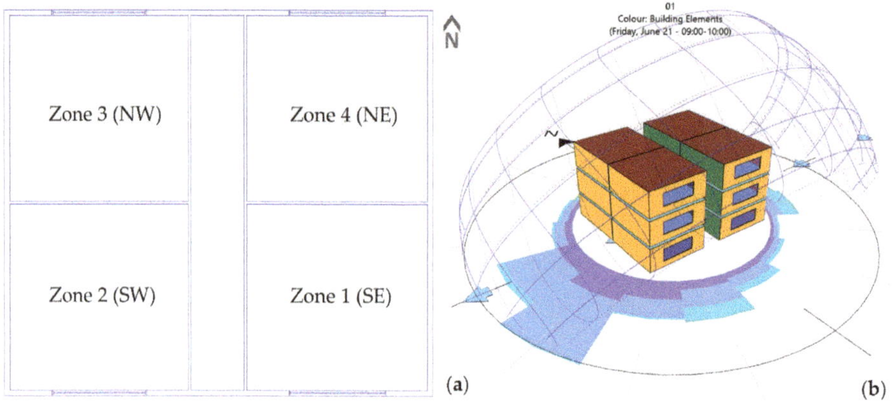

Figure 3. The office building (**a**) typical floor plan and (**b**) three-dimensional (3D) view in the case of a 10% window-to-floor (WFR) ratio with assigned north- and south-facing windows.

Table 4. Glazing material properties generated by TAS software [55].

Glass Type	Materials (Internal to External)	G Value	Light Transmittance	Emissivity Int./ext.	Conduct. $(W/m^2 \cdot {}^\circ C)$	U Value $(W/m^2 \cdot {}^\circ C)$	R Value $(m^2 \cdot {}^\circ C/W)$
Double glass	4 mm clear glass, 10 mm air gap, 4 mm clear glass	0.748	0.815	0.845	5.958	2.96	0.338

Table 5. Opaque construction materials and specifications generated by TAS software [55].

Construction	Materials (Internal to External)	Solar Absorptance	Emissivity Int./ext.	Conduct. $(W/m^2 \cdot {}^\circ C)$	U Value $(W/m^2 \cdot {}^\circ C)$	R Value $(m^2 \cdot {}^\circ C/W)$
External wall	Cement plaster 25 mm, clay hollow bricks 250 mm, cement plaster 25 mm	0.400	0.900	0.416	0.388	2.576
Internal wall	Cement plaster 25 mm, clay hollow bricks 100 mm, cement plaster 25 mm	0.400	0.900	0.745	0.661	1.512
Internal floor/ceiling	Concrete internal floor/ceiling 150 mm	0.650	0.900	7.533	3.303	0.303
Ground floor	Tiles 25 mm, mortar 50 mm, concrete 125 mm, aggregate 75 mm, soil 1000 mm	0.760	0.910	0.296	0.282	3.543
Roof	Cement plaster 25 mm, concrete 200 mm	0.650	0.900	2.027	1.507	0.663

3.4. Parameters of Thermal Simulations and Internal Conditions

The internal conditions were set as natural ventilation without any plant. Therefore, there were no active systems running for heating, cooling, or mechanical ventilation. The internal heat gain sources and coefficients are specified based on the TAS system parameters [55], as shown in Table 6. The Chartered Institution of Building Services Engineers (CIBSE) Guide A: Environmental Design [59] benchmark allowances were used to identify the values of internal heat gains, as summarised in Table 7.

The metabolic rate was predicted to be 1.2 met of 1.8 m^2 Du Bois area of an average adult doing sedentary office work, which indicates a heat release of 126 W/person [60].

To calculate the pollutant (i.e., CO_2) generation rate, ASHRAE fundamentals [61] and ASHRAE 62.1 standard [62] suggest that the CO_2 generation rate, for an average-sized adult performing sedentary office activities (1.2 met) is 0.0052 L/s (0.312 L/min). Referring to the range of 6 m^2 (open office) to 10 m^2 (single office) floor area per person required by office design guidelines and recommended area per person [57,61–63], each zone was designed to accommodate 6 people. Thus, the total CO_2 generation for a single zone will be 2.24 L/h/m^2. An amount of 7.5 L/s (15 cfm) per person of outdoor air can, therefore, dilute the polluted air. Natural ventilation through openable windows is the main conduit for the flow of air in and out. According to '*Tas Theory Manual*' [64], the wind pressure coefficients are defined in a way that the wind pressure on an aperture is:

$$p_w = \frac{c_w \rho v(h_b)^2}{2}. \tag{1}$$

where c_w is the wind pressure coefficient, ρ is the air density, and $v(h_b)$ is the wind speed at the building height (h_b).

All the parameters affecting wind pressure coefficient were based on the metrological weather data of Famagusta, as well as the terrain roughness was set to urban and cities category with terrain-dependent coefficients of exponent ($\alpha = 0.33$) and boundary layer thickness ($\delta = 460$). Finally, the occupancy schedule was set to weekdays (Monday to Friday) and office working hours only (09:00 to 17:00) for both internal conditions and aperture openings. Therefore, the total working days is 261 days and the total simulated hours is 2088 h. Generated by the North Cyprus metrological office and Famagusta weather station, the weather file data—in the format of TAS weather data (.twd)—of Famagusta was entered, which contains all the geographical data and variables for each hour of a year.

Table 6. The sources of internal heat gains and coefficient limitations based on TAS system parameters [55].

Internal Heat Gain Sources	Radiation Proportion	Coefficient
Lighting	0.3	0.490
Occupant	0.2	0.227
Equipment	0.1	0.372

Table 7. Inputs for internal gains based on the Chartered Institution of Building Services Engineers (CIBSE) Guide A benchmark allowances [55] and American Society of Heating, Refrigerating, and Air-Conditioning Engineers (ASHRAE) fundamentals [57].

Internal Gain/System Inputs	Unit	Value
Outside air	L/s/p	8.0
Metabolic rate	W/p	126.0
Infiltration	ach	0.3
Ventilation *	ach	0.0
Lighting gain	W/m^2	12.0
Occupancy sensible gain	W/m^2	8.0
Occupancy latent gain	W/m^2	5.0
Equipment sensible gain	W/m^2	18.0
Equipment latent gain	W/m^2	0.0
CO_2 pollutant generation	L/h/m^2	2.24

* Natural ventilation through windows (no mechanical ventilation).

3.5. Performance Criteria and Assessment Methods

Window design and natural ventilation are associated with many aspects of building indoor environmental quality (IEQ). The scope of this study involves an evaluation of the relationship between window design and natural ventilation in terms of CO_2 and thermal comfort performance. The following sections describe the assessment methods of these performance criteria.

3.5.1. Assessment of Carbon Dioxide (CO_2) Performance

High carbon dioxide concentration indoors can be an indicator of poor air circulation or under-ventilation. An indoor concentration greater than 1000 ppm of CO_2 is indicative of a potential indoor air quality problem [44]. CO_2 concentration below 1000 ppm usually indicates that the ventilation is adequate to deal with the normal products associated with human occupancy. In addition, the British and European standard BS EN 15251:2007 [65] categorises CO_2 levels above the outdoor concentration into four categories, as demonstrated in Table 8.

Due to the existence of a close relationship between CO_2 production and body odour, CO_2 level increases or decreases in relation to human metabolic activity. Since CO_2 is a good indicator of human metabolic activity, it could also be used as a tracer for other human-emitted bio-effluents. Moreover, CO_2 can be used to measure or control any per-person ventilation rate, regardless of the perceived level of bio-effluents or body odour in a given space. In fact, the 1000 ppm guideline for CO_2 recommended by the World Health Organisation [44] and used in ANSI/ASHRAE Standard 62.1 [62] is the equilibrium level for 15.0 cfm/person (7.0 L/s), assuming a 400 ppm outside CO_2 level. More recently, ANSI/ASHRAE Standard 62.1 indicated that comfort (odour) criteria are likely to be satisfied when the ventilation rate is set so that the 1000 ppm of CO_2 threshold is not exceeded [62]. Accordingly, the average duration (hour) of exposure to carbon dioxide concentration more than 1000 ppm per person per day can be measured [40].

Table 8. Building categories according to CO_2 levels above outdoor level based on British and European standard BS EN 15251 [65] and BS EN 13779 [66] standards.

Category	CO_2 Concentration (ppm) above Outdoor Air		The Accepted Limit for Famagusta (Outdoor CO_2 of 400 ppm)
	Typical Range	**Default Value**	
I	≤400	350	750
II	400–600	500	900
III	600–1000	800	1200
IV	>1000	1200	1600

3.5.2. Thermal Comfort Assessment Using an Adaptive Model

In the 1970s, an adaptive comfort theory challenged the steady-state comfort theory, which suggested that comfort was time-dependent considering human thermal adaptation (i.e., behavioural, physiological, and psychological) to their environment over time. Thus, the building occupants might accept conditions that would otherwise have been predicted to be unsatisfactory for the PMV model [67], specifically in the hot conditions of naturally ventilated buildings [68]. The model hypothesis is that contextual factors influence building residents' preferences and thermal expectations [69,70]. The concept of the adaptive comfort model is that outdoor climate impacts indoor comfort as occupants can adapt to different conditions throughout different times of the year. The results of field studies revealed that users of naturally ventilated buildings typically accept a wider range of temperatures than those in air-conditioned buildings as their preferred temperature depends on outdoor conditions [2,71]. The model works efficiently in an environment where the monthly mean temperature stays above 10 °C and below 33.5 °C, which corresponds to the weather conditions of Famagusta, North Cyprus.

Similar to the acceptability limits of 80% and 90% defined by the ASHRAE 55 standard [60], the British and European standard of BS EN 15251:2007 [64] introduced a similar categorisation using Equation (2) while accepting slightly higher degrees than the American standard. It was proposed that an exponentially weighted outdoor running mean temperature could account for this time-dependency. Therefore, the BS EN15251:2007 standard defines the exponentially weighted running mean temperature T_{rm} for any given day through Equation (3), which was originally developed by Nicol and Humphreys [72].

$$T_{comf} = 0.33 \cdot T_{rm} + 18.8, \tag{2}$$

$$T_{rm} = (1 - \alpha)T_{od-1} + \alpha T_{od-2} + a^2 T_{od-3} + a^3 T_{od-4} \ldots), \tag{3}$$

where T_{comf} is the indoor comfortable operative temperature (°C) and T_{rm} is the exponentially weighted running mean temperature (°C), α is a constant between 0 and 1 and T_{od-1} is yesterday's daily mean outdoor temperature, the day before (T_{od-2}), the day before that (T_{od-3}), and so on.

The temperatures become less significant as time progresses, with the speed of decay depending on the value of the constant α. The lower the value of α, the less significant the weighting of past temperatures. Moreover, the equation's developers suggested $\alpha = 0.8$ as an appropriate value according to their SCAT database [72]. Table 9 explains that the standard defines three categories of comfort ranges for different expectations. Moreover, occupants accept temperatures within the comfort ranges as comfortable, and consider temperatures outside of the upper and lower limits too hot and too cold, respectively.

Table 9. Thermal comfort categories and acceptable ranges based on the European adaptive model [65].

Categories	Acceptable Comfort Range (°C)		Expectations
	Upper Limit	**Lower Limit**	
I	$T_{comf} = 0.33 \cdot T_{rm} + 18.8 + 2$	$T_{comf} = 0.33 \cdot T_{rm} + 18.8 - 2$	High
II	$T_{comf} = 0.33 \cdot T_{rm} + 18.8 + 3$	$T_{comf} = 0.33 \cdot T_{rm} + 18.8 - 3$	Normal
III	$T_{comf} = 0.33 \cdot T_{rm} + 18.8 + 4$	$T_{comf} = 0.33 \cdot T_{rm} + 18.8 - 4$	Moderate

4. Simulation Results and Analysis

The main findings of this study can be divided into two parts. First, the results of the effect of window design and natural ventilation on CO_2 concentration are presented and analysed. Second, the results of thermal comfort performance using an adaptive model are provided and analysed, followed by the discussion of main findings and conclusions drawn from the experimental results in the followed sections.

4.1. Effect of Window Design and Natural Ventilation on CO_2 Concentration

The measurements of indoor carbon dioxide levels were initiated with a 10% window-to-floor ratio as the minimum window area required by the building guidelines in North Cyprus. The window opening ratios ranged from fully closed to fully opened windows, while the window orientations were south-, east-, north-, and west-facing windows, divided into four thermal zones on each floor. To explore the impact of single-side and cross-flow ventilation, various window sizes (i.e., 10%, 25%, and 50% WFR), openings (i.e., 0%, 10%, 25%, 50%, 75%, and 100%), and different window orientations are applied to all zones in the ground, first, and second floor. Initially, a fully closed (0% open) window corresponds to a situation where neither openable windows nor mechanical ventilation is provided. In the free-running period, however, this is not a practical scenario because window-based natural ventilation might be the only means to modify indoor conditions in terms of air quality and thermal comfort. In air-conditioned buildings, the case represents having no adequate mechanical or mixed-mode ventilation. The CO_2 amount of difference for adjacent zones having the same window orientation and design was less than 2 ppm; therefore, the results of the similarly performing zones were excluded.

4.1.1. Results of Single-Sided Natural Ventilation

The indoor carbon dioxide level exceeded the ANSI/ASHRAE 62.1 and WHO recommended threshold (1000 ppm), in all the cases of different window sizes and orientations, when the windows are fully closed. Figure 4 illustrates the percentages of office hours where the CO_2 level is below the WHO threshold (1000 ppm) for first floor zones having a 10% WFR with single-side ventilation. Table 10 summarises the number of annual occupancy hours appearing in each CO_2 category based on the BS EN 15251:2007 standard. When the 10% (or 25%) window-to-floor ratio is closed at all times, none of the zones provides any office working hours that the CO_2 concentration appears under the category I (<750 ppm) and II (750–900 ppm). When the 10% WFR is opened by 10% during occupancy hours (08:00–17:00), considerable improvement can be seen for all window orientations.

Table 10. The number of office occupancy hours appearing in the CO_2 categories of BS EN 15251:2007 standard for first floor zones.

WFR (%)	Ventilation Strategy	Opening Ratio (%)	CO_2 Categories	Window Orientations			
				S Win	E Win	N Win	W Win
10% 25% 50%	Single-side or Cross-flow	Closed	I	0	0	0	0
			II	0	0	0	0
			III	261	261	261	261
			IV	1827	1827	1827	1827
10%	Single-side	10% open	I	515	444	269	314
			II	1239	1498	1561	1698
			III	282	134	234	7
			IV	52	12	24	4
		25% open	I	1891	2024	1980	2059
			II	153	50	93	25
			III	44	13	15	4
		50% open	I	2044	2083	2070	2085
			II	39	5	17	3
			III	5	0	1	0
		75% open	I	2085	2088	2087	2088
			II	3	0	1	0
			III	0	0	0	0
		Fully open	I	2088	2088	2088	2088
25%	Single-side	10%open	I	2039	2081	2043	2080
			II	37	7	40	8
			III	12	0	5	0
		25%open	I	2088	2088	2088	2088

* Blue colour and orange colour indicate the most and least effective window orientations respectively.

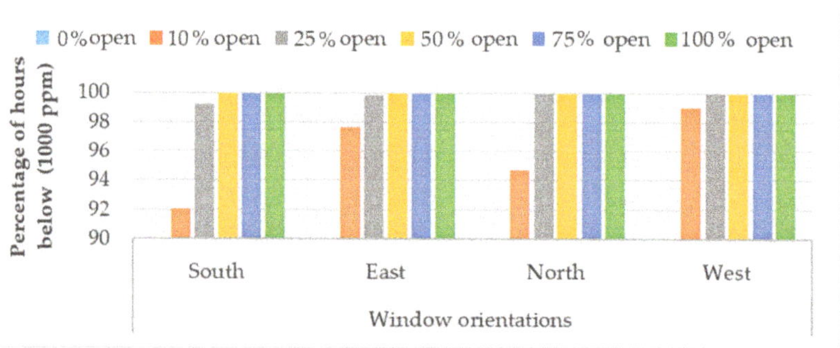

Figure 4. Percentages of office occupancy hours that the CO_2 level is below the WHO threshold (1000 ppm) for first floor windows in the case of a 10% WFR with single-side ventilation.

When single-side ventilation of a 10% WFR is opened by 10%, the east-facing windows provide more hours within the category I (<750 ppm) for the ground- and second-floor zones, followed by south-facing windows. Zone 1 (SE) had the most efficient natural ventilation performance that dilutes the maximum amount of CO_2 and provides 837 and 790 h (out of 2088 annual occupancy hours) of category I through east- and south-facing windows, respectively. Conversely, most of the category II (750–900 ppm) hours can be seen on the second-floor zones, which ranges between 1514–1770 h in zone 9 (south- and east-facing windows) and 1727–1785 h in zone 11 (north- and west-facing windows). In addition, the zones with south- and east-oriented windows have not recorded any hours in either category III or IV on the second floor. These results were also approximately noticed in the eastern and western windows of the ground floor.

Overall, in the cases of single-sided natural ventilation, the west-facing windows provided the maximum number of annual occupancy hours within category II when the 10% WFR is 10% opened, followed by east-facing windows. Moreover, increasing the ratio of window openings (e.g., equal to or greater than 25%) improves the natural ventilation performance of western and eastern windows, while the south-oriented windows become the least effective window orientation. The performance of different window orientations is convergent in the greater opening ratios, such as 75% window opening and onward, with approximately providing all the annual occupancy hours inside category I. The single-side natural ventilation performance of a 10% WFR having 50% of the area opened is similar to a 25% WFR with a 10% window opening. Furthermore, if a 25% WFR is opened by 25%, all the office working hours appear inside Category I.

Figures 5 and 6 demonstrate the level of CO_2 concentration in warm and cool periods for different window orientations and opening ratios in the case of single-side ventilation for 25% and 10% WFR, respectively. The findings presented in Figures 4 and 5 explain that ground floor zones have a maximum CO_2 level when the windows are fully closed. While, first floor zones have the highest CO_2 concentration when the windows are opened by any opening ratios, particularly the south-facing window in the summer (855 ppm) and north-facing window in the winter (845 ppm). The performance of south- and east-facing windows are noticeably higher than north- and west-facing windows on each floor. In the summer months, all the window orientations perform better than the winter period, except south-facing windows, which show the opposite results. A window opening of 25% provided category I for any window orientation, where the range was between 580–685 ppm in both the warm and cool periods. The various window opening ratios for a 25% WFR show an identical pattern to the 10% WFR with the only difference being that a lesser CO_2 concentration was achieved.

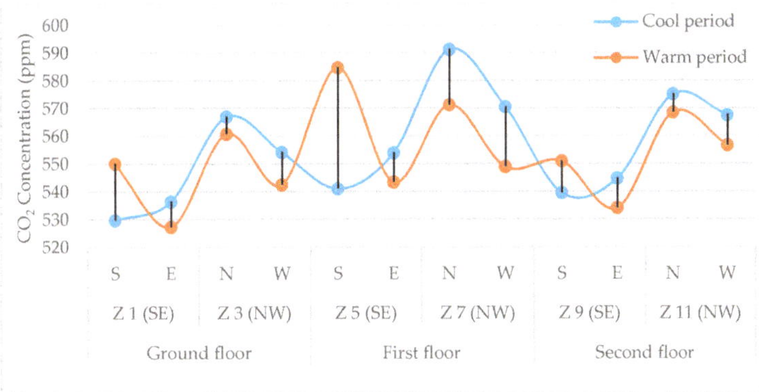

Figure 5. The CO_2 concentration (ppm) in cool and warm months in the case of single-side ventilation with a 25% WFR and 10% opened windows.

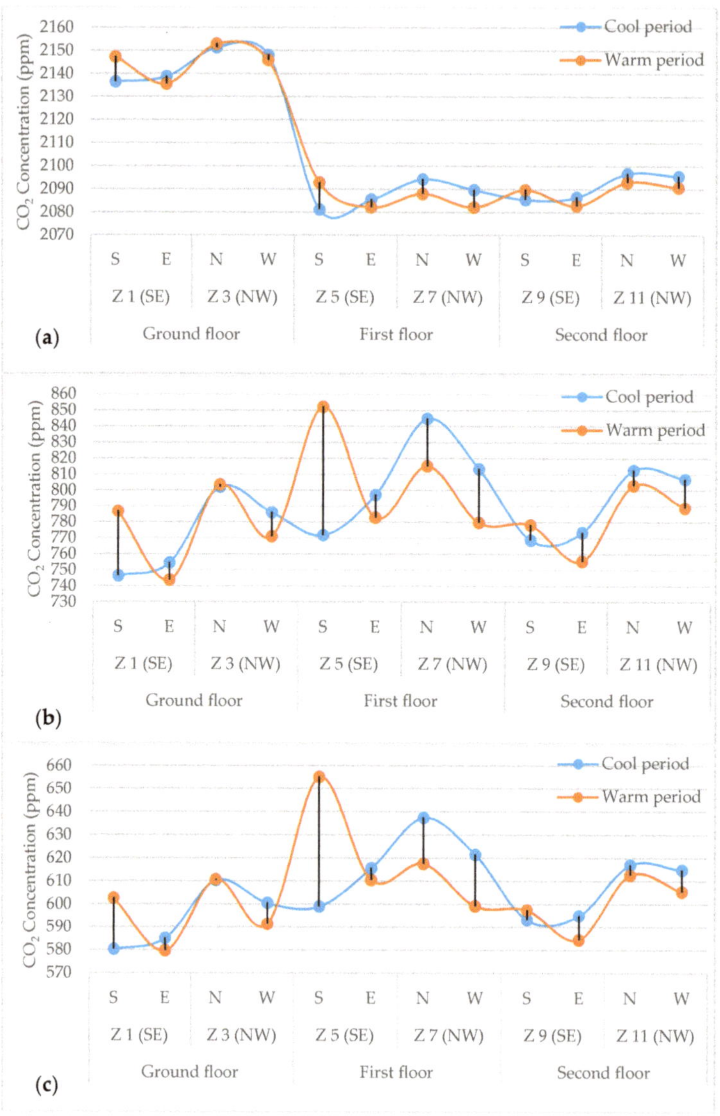

Figure 6. The CO_2 concentration (ppm) in cool and warm months in the case of single-side ventilation with a 10% WFR and (**a**) closed windows, (**b**) 10%, and (**c**) quarter opened windows.

4.1.2. Results of Cross-Flow Natural Ventilation

A cross-flow ventilation strategy was assigned to 10% and 50% (fully glazed wall) WFRs, for which significant improvements can be noticed compared to single-side ventilation scenarios. Table 11 summarises the number of annual occupancy hours appearing in each CO_2 category based on the BS EN 15251:2007 standard in the case of cross-ventilation. For a 10% WFR, an opening of 10% can ensure most of the office occupancy hours inside category I and II. This fracture of opening in the case of fully glazed wall offers all the 2088 annual office hours within the category I. This objective can be achieved with 25% window opening in the case of a 10% WFR. Overall, the second floor zones showed better results in its natural ventilation potentials. Taking the second floor as the ideal natural ventilation performance, the most effective window orientations were a combination of the south- and east-facing windows (Zone 9: 1901 h of category I), followed by north- and east-facing windows (Zone 12: 1710 h of category I). However, the least performing window combination for the cross-ventilation method was north- and west-oriented windows (Zone 11: 1487 h of category I).

Finally, Figures 7 and 8 display the CO_2 level in warm and cool months for different zones in the case of cross-flow ventilation for a 10% WFR and fully glazed external wall, respectively. In both window sizes, a 10% window opening can place all the annual occupancy hours inside category I for each zone. Overall, opening 10% of the windows can lower the CO_2 level to under 720 ppm for a 10% WFR and 470 ppm for a fully glazed wall in both winter and summer. In the cool and warm periods, second floor zones record less CO_2 concentration than first and ground floor. Noticeably, the zones with a combination of south and east windows for cross ventilation are more effective than any other window orientations in both the summer and winter months. However, the zones with cross-flow ventilation from the north- and west-facing windows perform less well than other window orientations, particularly in the warm period. Nevertheless, in the cool period, window combinations for cross ventilation show similar results, except a combination of south and east window, which recorded lower CO_2 levels.

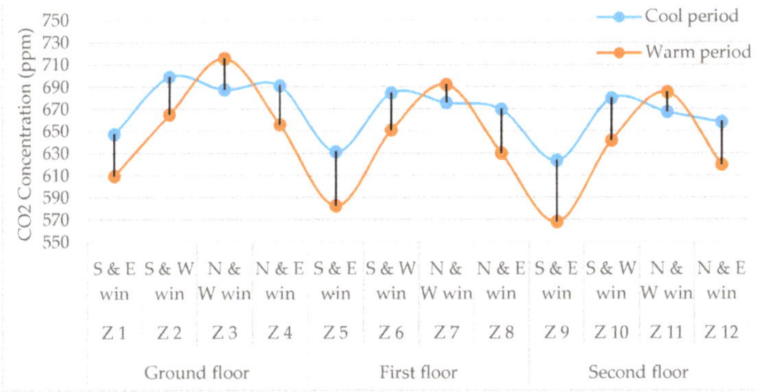

Figure 7. The CO_2 concentration (ppm) in cool and warm months in the case of cross-flow ventilation with a 10% WFR and 10% opened windows.

Figure 8. The CO_2 concentration (ppm) in cool and warm months in the case of cross-flow ventilation with a 50% WFR and 10% opened windows.

Table 11. The number of annual occupancy hours appearing in the CO$_2$ categories based on BS EN 15251:2007 standard in the case of cross-flow ventilation.

WFR (%)	Ventilation Strategy	Opening Ratio (%)	CO$_2$ Categories	Ground Floor Zones/Windows				First Floor Zones/Windows				Second Floor Zones/Windows			
				Z 1 (SE)	Z 2 (SW)	Z 3 (NW)	Z 4 (NE)	Z 5 (SE)	Z 6 (SW)	Z 7 (NW)	Z 8 (NE)	Z 9 (SE)	Z 10 (SW)	Z 11 (NW)	Z 12 (NE)
				S + E Win	S + W Win	N + W Win	N + E Win	S + E Win	S + W Win	N + W Win	N + E Win	S + E Win	S + W Win	N + W Win	N + E Win
10%	Cross-flow	10% open	I	1849	1492	1280	1505	1904	1533	1421	1653	1901	1549	1487	1710
			II	239	596	804	583	184	555	662	433	187	539	592	377
			III	0	0	4	0	0	0	5	2	0	0	9	1
		25% open	I	2088	2088	2088	2088	2088	2088	2088	2088	2088	2088	2088	2088
50%	Cross-flow	10% open	I	2088	2088	2088	2088	2088	2088	2088	2088	2088	2088	2088	2088

* Blue colour and orange colour indicate the most and least effective window orientations respectively.

4.2. Results of Adaptive Thermal Comfort

4.2.1. Findings of Single-Sided Natural Ventilation Using an Adaptive Model

The results of single-side natural ventilation show that when the zones are assigned the minimum window-to-floor ratio (10%), different performances can be noticed with respect to various window orientations, opening ratios, and floor locations, as reported in Table 12. Firstly, in the case of fully closed windows, the zones provide minimal hours that are comfortable based on the adaptive comfort categories of the BS EN 15251:2007 standard, noting second-floor zones perform better compared to the first floor and ground floor zones, respectively. When a 10% window area was opened, the south-facing windows produce more thermally uncomfortable indoor environments than the other window orientations, followed by eastern windows. Conversely, north- and west-facing windows provide more hours of adaptive comfort, respectively.

Nevertheless, the results of the quarter, half, three-quarter, and full window openings display contradictory window and natural ventilation performances compared to previous scenarios. When a quarter of the 10% WFR was opened, south-facing windows on the second floor achieved the highest number of thermal comfort hours inside Category I and II of the European adaptive comfort model, specifically 611 and 858 h, respectively, out of 2088 annual office working hours. While the other window orientations provided a convergent number of comfortable hours on this floor, which ranged between 555 to 573 h in Category I and 783 to 807 in Category II, it is worth mentioning that the east window represents the least efficient case. On the other hand, southern windows are less effective on the ground and first floors when only a quarter of the window area is opened during office working hours. West- and north-facing windows offer more hours that are comfortable than eastern windows.

In contrast to the 10% and 25% window openings, the southern and eastern windows can perform better than west- and north-facing windows if half, three-quarter, or the full area of the windows is kept open during office hours, regardless of whether it is located on the ground, first, or the second floor. Moreover, through this particular opening ratio, ground floor windows are more efficient than the first- and second-floor windows for all window orientations. Opening 50% of the southern window in zone 1 (SE) provides 918 and 1045 h, zone 5 (SE) contributes to 825 and 987 h, and zone 9 (SE) allocates 803 and 985 h in category I and category II of the adaptive model, respectively.

In the case of a 25% window-to-floor ratio, as presented in Table 13, north- and east-oriented windows performed slightly better only when 10% of the window area was open, compared to the same scenario of 10% WFR. Conversely, northern and western window orientations presented a less effective performance in all window-opening ratios on each floor location. In contrast to the 10% WFR case, increasing the opened portion for south- and east-facing windows offer more hours in category I and II on each floor. The other window orientations reduce their efficiency with a larger window opening area regardless of the floor location. Overall, the order of most and least efficient window orientations is almost the same as to the 10% WFR. Figures 9 and 10 illustrate the effect of window design on the thermal comfort performance of a naturally ventilated office building during cool and warm periods. Both 10% and 25% window-to-floor ratios manifest comparable results with the domination of too warm percentages in the summer months nearly in all window-opening ratios. By looking at a 10% window opening in both the window sizes, one can notice that approximately all window orientations are considered too warm during the summer months. Furthermore, in the cool period, south-facing windows represent the worst scenarios when the windows are closed, particularly on the ground and the first floor, with comfort around only 30% of the time, while 70% is considered too warm as a reason of overheating, mostly by internal gains, as well as solar radiation. A 10% window opening offers the least amount of hours that are considered comfortable according to category III of the European adaptive model, which is less than 10% during the warm period. Nevertheless, a slightly better performance can be seen in the case of 25% WFR.

Table 12. The number of comfort hours for adaptive model categories based on BS EN 15251:2007 standard in the case of 10% WFR with single-side ventilation.

| Window Opening Ratio (%) | Adaptive Comfort Categories | Ground Floor/Windows | | | | | | | | Second Floor/Windows | | | |
| | | Z 1 (SE) | | Z 3 (NW) | | Z 5 (SE) | | Z 7 (NW) | | Z 9 (SE) | | Z 11 (NW) | |
		S Win	E Win	N Win	W Win	S Win	E Win	N Win	W Win	S Win	E Win	N Win	W Win
0%	Category I	0	0	285	161	0	8	299	197	9	92	357	276
	Category II	0	12	464	312	1	44	457	327	26	155	514	407
	Category III	0	77	675	468	5	119	633	469	61	222	679	541
10%	Category I	36	223	804	622	47	238	718	556	177	378	617	532
	Category II	92	380	961	818	127	371	896	775	369	547	836	726
	Category III	336	548	1049	919	327	521	1011	888	601	703	977	871
25%	Category I	516	606	603	637	435	538	613	620	611	555	573	565
	Category II	843	745	978	924	744	694	903	884	858	783	805	807
	Category III	1017	883	1217	1099	955	811	1127	1052	990	926	1037	996
50%	Category I	918	607	497	529	825	577	499	543	803	570	459	507
	Category II	1045	918	784	840	987	828	803	819	985	810	754	767
	Category III	1147	1067	1184	1128	1097	987	1109	1077	1097	1041	1015	1001
75%	Category I	902	620	448	498	855	574	468	501	764	572	437	480
	Category II	1088	887	741	786	1040	828	735	783	1003	787	698	734
	Category III	1217	1119	1109	1081	1140	1044	1067	1059	1143	1042	972	1001
100%	Category I	866	576	407	472	837	574	431	475	727	511	416	455
	Category II	1092	857	719	746	1041	825	697	752	986	792	666	706
	Category III	1282	1129	1077	1060	1185	1041	1023	1033	1181	1012	953	982

Note: columns Z 5 (SE) and Z 7 (NW) belong to the "First Floor/Windows" group.

* Blue colour and orange colour indicate the most and least effective window orientations respectively.

Table 13. The number of comfort hours for adaptive model categories based on BS EN 15251:2007 standard in the case of 25% WFR with single-side ventilation.

Window Opening Ratio (%)	Adaptive Comfort Categories	Ground Floor/Windows								First Floor/Windows								Second Floor/Windows							
		Z 1 (SE)		Z 3 (NW)						Z 5 (SE)		Z 7 (NW)						Z 9 (SE)		Z 11 (NW)					
		S Win	E Win	N Win	W Win					S Win	E Win	N Win	W Win					S Win	E Win	N Win	W Win				
0%	Category I	0	0	307	99					0	0	317	127					1	10	354	178				
	Category II	0	0	491	181					0	0	474	215					2	35	511	281				
	Category III	0	0	651	303					0	5	625	326					7	83	654	388				
10%	Category I	65	294	571	566					65	270	584	532					137	393	554	511				
	Category II	154	426	883	799					137	380	849	758					270	553	774	715				
	Category III	301	564	1149	948					259	513	1084	909					440	693	1007	897				
25%	Category I	384	516	469	538					310	450	461	523					434	507	433	509				
	Category II	642	703	730	781					549	635	723	760					661	707	701	725				
	Category III	837	829	1052	1031					770	770	1021	988					870	898	956	931				
50%	Category I	675	549	407	478					581	514	421	483					644	522	408	457				
	Category II	885	789	663	759					827	732	663	755					884	762	636	719				
	Category III	1057	965	968	1007					993	885	957	988					1062	963	917	952				
75%	Category I	765	544	391	455					699	526	391	466					685	518	396	443				
	Category II	966	794	653	734					910	749	635	733					949	770	627	703				
	Category III	1126	1006	958	1001					1069	938	920	985					1093	985	885	948				
100%	Category I	785	552	385	465					746	527	378	461					705	517	387	434				
	Category II	999	802	640	705					950	765	628	721					972	783	631	691				
	Category III	1155	1024	965	989					1096	967	911	974					1123	994	867	942				

* Blue colour and orange colour indicate the most and least effective window orientations respectively.

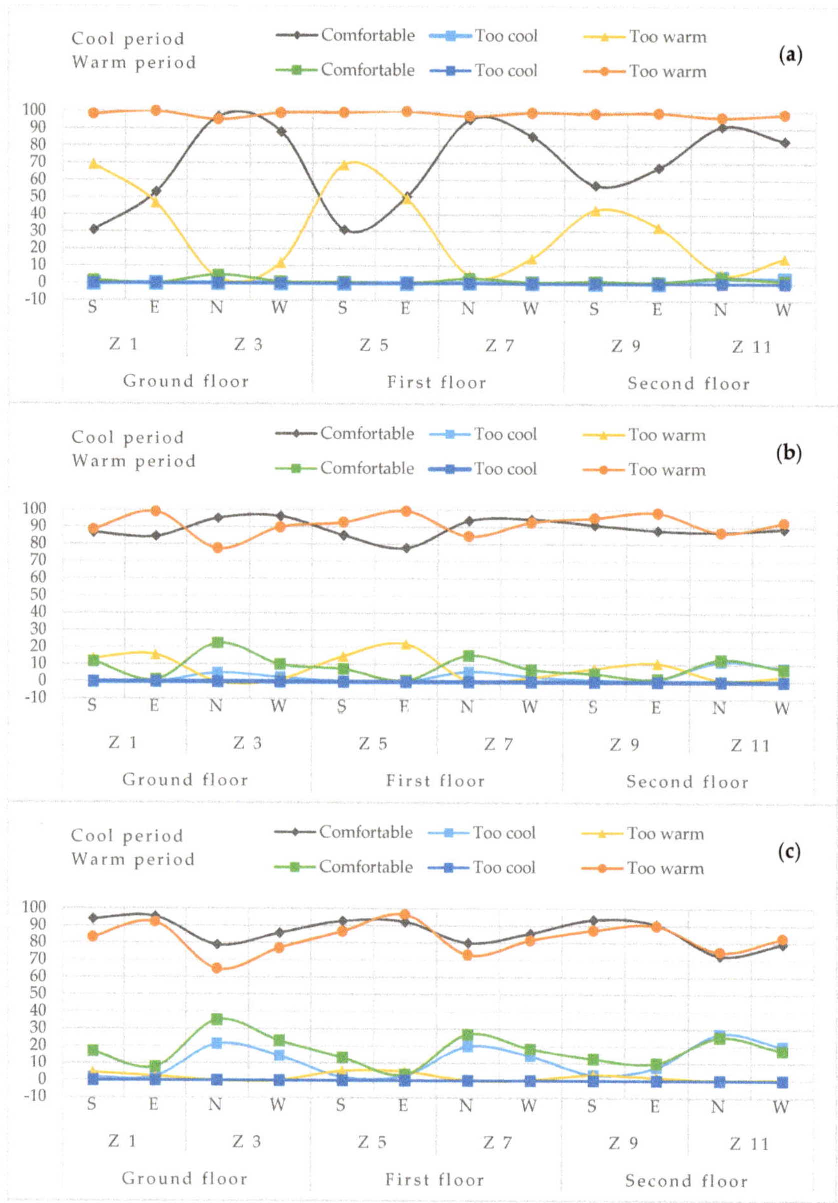

Figure 9. Percentages of thermal sensation in cool and warm months based on category III of the European adaptive model in the case of a 10% WFR for (**a**) 10%, (**b**) quarter, and (**c**) half-opened windows.

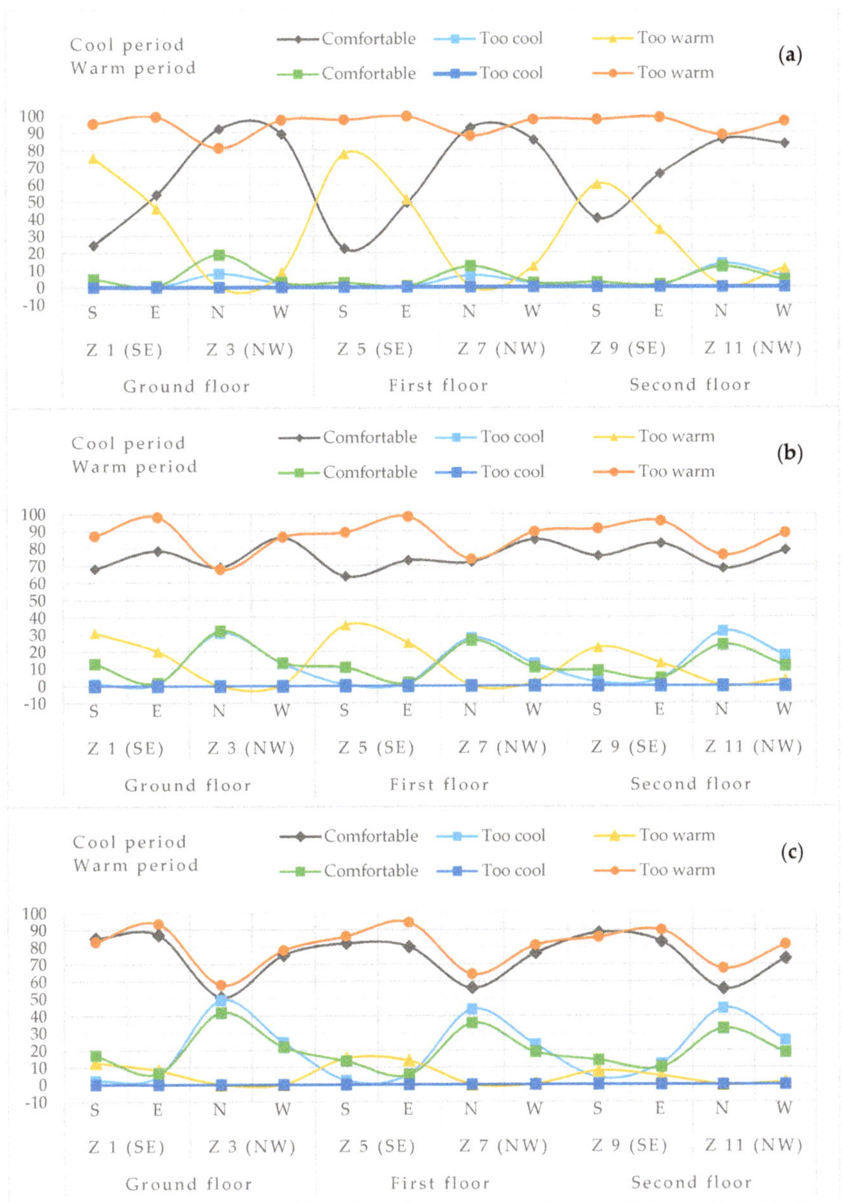

Figure 10. Percentages of thermal sensation in cool and warm months based on category III of the European adaptive model in the case of a 25% WFR for (**a**) 10%, (**b**) quarter, and (**c**) half-opened windows.

When opening quarter of the window area, nearly all window orientations perform better than the 10% window opening in both seasons, noting that the eastern windows are less effective than other window directions. The case of 25% WFR slightly improves thermal performance in the warm period but reduces the number of acceptable hours in the winter through the increase in cooler sensations.

The half window opening enhances indoor thermal comfort in the warm period while simultaneously decreasing the number of hours that appear in the acceptable range of category III of adaptive comfort.

4.2.2. Findings of Cross-Flow Natural Ventilation Using Adaptive Model

Tables 14 and 15 outline the number of office occupancy hours appearing in the European adaptive comfort categories in the case of 10% and 50% WFR with cross-ventilation. In the case of a fully glazed external wall, cross-ventilation improves indoor thermal comfort when increasing window opening ratios. When opening 10% of the window area, the zones that have a window combination of the north- and west-facing windows for the 10% WFR and east-facing windows for the fully glazed wall display better results. Conversely, increasing the window opening from 25% to 100% can gradually provide a greater number of comfortable hours for the zones with a window combination of the south- and east-oriented windows and a 10% WFR as well as south- and west-facing windows for the fully glazed wall. Such increments in window opening confirm that cross ventilation form north- and west-oriented windows have the least efficient natural ventilation performance compared to other window orientations in any window size and opening ratio.

Cross ventilation through a 10% WFR with various window orientations, openings, and floor locations are presented in Figure 11. First, a 10% window opening is least effective in the overheating period but performs better than other scenarios in the winter months. About 30% to 40% of the occupancy hours were thermally acceptable when a half area of the window was opened in the warm period. Fully opened windows raise this percentage, with 50% of the office occupancy time being comfortable. In general, having cross ventilation through a combination of the north- and west-facing windows is the most effective case in the warm period, in nearly all opening scenarios. Although this situation could also be observed in the cool period if only 10% of the window area is opened. A scenario of having cross ventilation from the south- and east-oriented windows performed better in the winter months and at opening ratios larger than 10%.

The sun's intense rays reduced the effectiveness of cross-ventilation in the case of fully glazed external windows, as illustrated in Figure 12. Unshaded large glass surfaces can receive a significant amount of harmful solar radiation, which results in space overheating in the summer months. It was observed that a 10% window opening led to more than 50% too warm condition even in the winter for the windows that receive a greater amount of solar radiation (i.e., south- and east-oriented). Despite the fact that greater window openings can cool down the indoor temperature, the too cool condition raises in the zones with the north-and west-facing windows when the windows are kept open during the occupancy hours in the cool period. The occurrences of zone overheating stayed similar to testing various window-opening scenarios. Therefore, protecting windows or solar control is highly recommended if better thermal comfort conditions are desired in naturally ventilated office buildings in the Mediterranean climate.

Table 14. The number of comfort hours for adaptive model categories based on BS EN 15251:2007 standard in the case of 10% WFR with cross-ventilation strategy.

WFR (%)	Ventilation Strategy	Opening Ratio (%)	CO_2 Categories	Ground Floor Zones/Windows				First Floor Zones/Windows				Second Floor Zones/Windows			
				Z 1 (SE)	Z 2 (SW)	Z 3 (NW)	Z 4 (NE)	Z 5 (SE)	Z 6 (SW)	Z 7 (NW)	Z 8 (NE)	Z 9 (SE)	Z10 (SW)	Z11 (NW)	Z12 (NE)
				S + E Win	S + W Win	N + W Win	N + E Win	S + E Win	S + W Win	N + W Win	N + E Win	S + E Win	S + W Win	N + W Win	N + E Win
10%	Cross-flow	10% open	I	409	494	676	666	443	500	638	595	614	600	554	575
			II	646	807	932	864	670	781	873	839	819	835	793	823
			III	869	984	1066	1031	897	969	1019	1021	991	956	957	1019
		25% open	I	801	855	479	639	787	790	473	601	706	677	450	543
			II	1002	1029	774	938	989	974	742	893	942	902	688	809
			III	1169	1105	1089	1172	1132	1121	1029	1088	1133	1085	956	1027
		50% open	I	660	678	412	537	637	629	404	525	550	539	375	476
			II	1022	1012	671	783	991	940	649	755	887	848	633	701
			III	1256	1147	996	1091	1211	1209	969	1039	1158	1127	917	983
		75% open	I	572	520	388	473	556	496	385	467	476	458	377	449
			II	949	951	648	722	907	919	635	701	796	824	619	677
			III	1267	1191	1023	1079	1206	1213	935	1005	1137	1111	902	948
		Fully open	I	525	466	392	421	501	451	374	432	457	422	381	408
			II	876	880	649	704	841	850	618	686	758	786	609	659
			III	1290	1206	1018	1080	1198	1213	960	1009	1124	1110	910	958

* Blue colour and orange colour indicate the most and least effective window orientations respectively.

Table 15. The number of comfort hours for adaptive model categories based on BS EN 15251:2007 standard in the case of 50% WFR with cross-ventilation strategy.

WFR (%)	Ventilation Strategy	Opening Ratio (%)	CO$_2$ Categories	Ground Floor Zones/Windows				First Floor Zones/Windows				Second Floor Zones/Windows			
				Z 1 (SE)	Z 2 (SW)	Z 3 (NW)	Z 4 (NE)	Z 5 (SE)	Z 6 (SW)	Z 7 (NW)	Z 8 (NE)	Z 9 (SE)	Z 10 (SW)	Z 11 (NW)	Z 12 (NE)
				S + E Win	S + W Win	N + W Win	N + E Win	S + E Win	S + W Win	N + W Win	N + E Win	S + E Win	S + W Win	N + W Win	N + E Win
50%	Cross-flow	10% open	I	212	295	443	480	221	304	443	472	352	364	425	488
			II	324	460	714	721	339	465	681	689	500	536	653	713
			III	462	649	950	925	483	655	917	908	652	747	887	821
		25% open	I	424	539	431	503	423	525	423	490	484	536	405	485
			II	611	813	649	741	600	771	636	730	692	787	628	707
			III	772	887	931	986	770	984	913	950	897	976	876	929
		50% open	I	511	681	419	485	488	655	411	483	505	631	398	443
			II	722	914	631	740	703	886	638	722	775	859	636	703
			III	935	952	909	972	918	1061	894	944	983	1029	871	915
		75% open	I	544	719	415	469	516	676	402	463	535	640	392	451
			II	784	940	645	730	757	914	626	718	799	889	628	701
			III	983	971	880	958	963	1079	877	928	1014	1041	860	907
		Fully open	I	557	722	417	481	544	692	399	466	542	637	392	448
			II	812	955	631	727	788	918	627	716	817	894	614	693
			III	1015	974	869	945	993	1088	857	932	1027	1044	857	917

* Blue colour and orange colour indicate the most and least effective window orientations respectively.

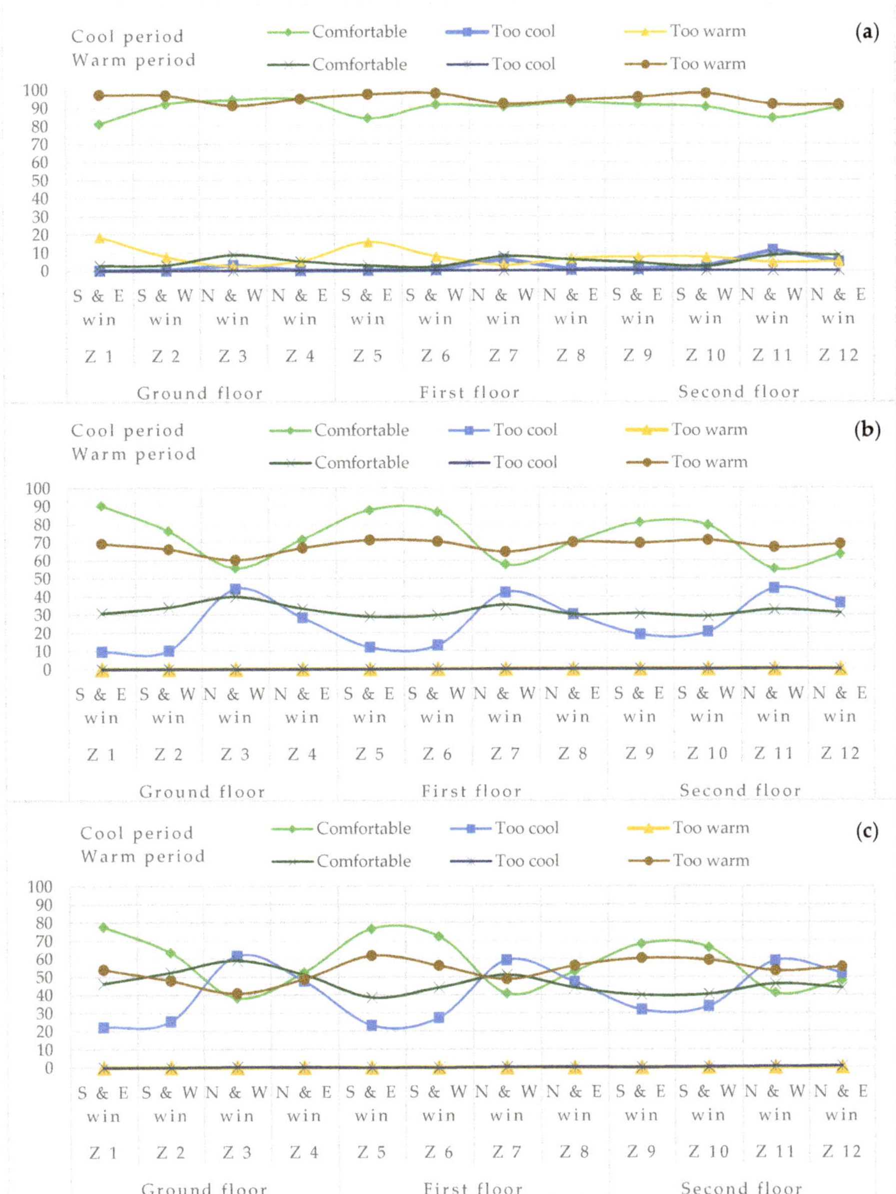

Figure 11. Percentages of thermal sensation in cool and warm months based on category III of the European adaptive model in the case of a 10% WFR with cross-ventilation for (**a**) 10%, (**b**) half, and (**c**) fully opened windows.

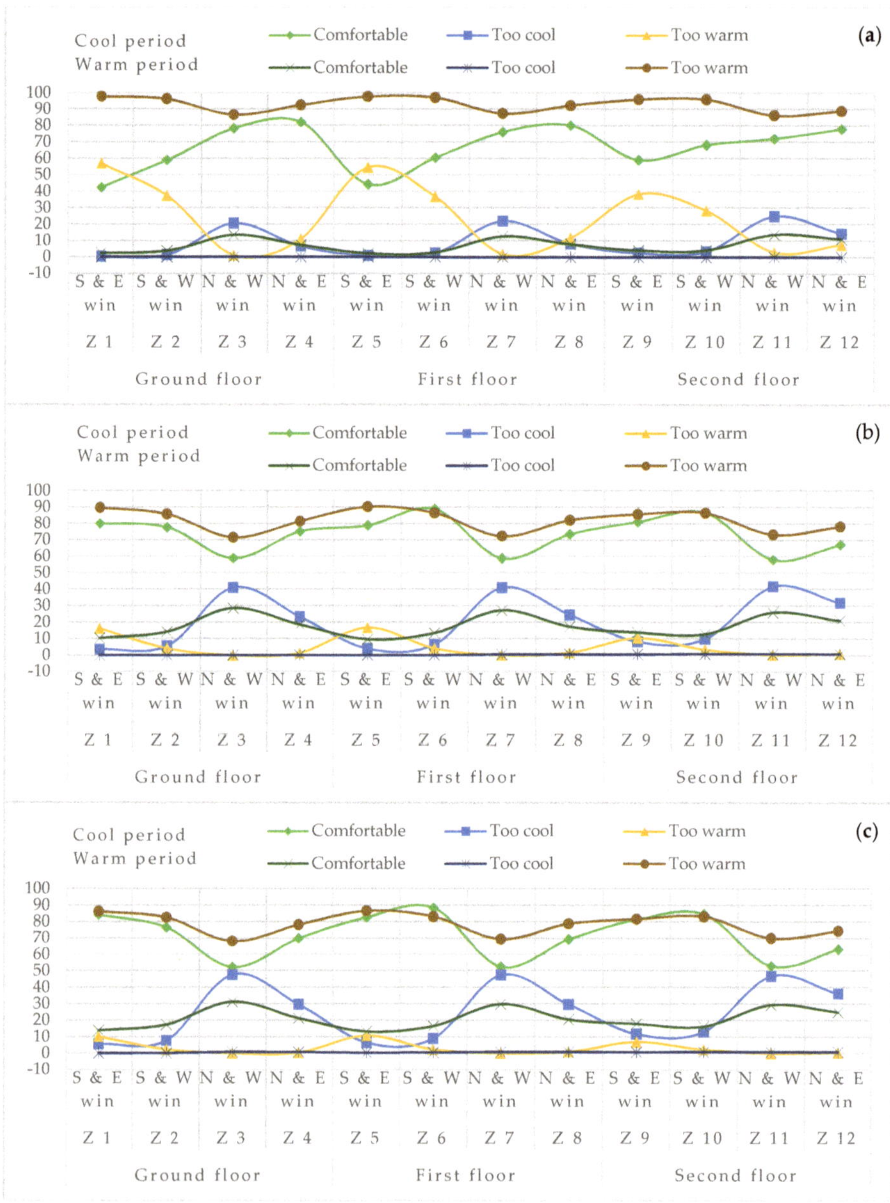

Figure 12. Percentages of thermal sensation in cool and warm months based on category III of the European adaptive model in the case of a fully glazed wall with cross-ventilation for (**a**) 10%, (**b**) half, and (**c**) fully opened windows.

5. Discussion and Concluding Remarks

5.1. Window and Natural Ventilation Performance in Terms of "Indoor CO_2 Level and Thermal Comfort"

Opening a window is a common and simple way of using natural ventilation to provide fresh air and cool the internal spaces of a building, but the airflow that occurs in this process is rather complicated due to the involvement of several parameters. The level of airspeed, wind direction, the temperature difference between inside and outside, pressure variations, and turbulence characteristics determine the amount of air coming through the openings. From an architectural point of view, the amount of airflow also depends on the size, orientation, location, fracture of opening, and type of window. Single-sided natural ventilation can become more complex compared to cross-flows by reason of involving both wind and thermal effects at the same time. In single-sided ventilation, the airflow through openings is mainly driven by the turbulence in the wind, in which space blocks the prevailing wind [4].

The results of this study indicate that, in the case of closed windows of any window size, location, or orientation, an average CO_2 concentration exceeding 2000 ppm can lead to various symptoms, and occupants are more likely to complain of headache, fatigue, and tiredness. In the free-running period, the window opening is a fundamental method of ventilation and air conditioning; thus, occupants use windows and other physiological adaptation mechanisms to maintain indoor air and thermal conditions. Therefore, closing windows is not acceptable neither for indoor air nor for thermal comfort conditions, even in the winter months. Moreover, in all the window orientations, first-floor zones recorded the worst ventilation performance in terms of CO_2 contamination as a reason for occurrence possible wind turbulence.

Table 16 presents the most and least effective window orientations, in terms of providing a maximum number of hours within category I CO_2 concentration based on the BS EN 15251:2007 standard, against different ventilation strategies, window sizes, and opening ratios. In the case of single-sided ventilation, the west- and east-facing windows provided more hours inside category I and II, while the south-facing windows represented the least effective orientation. These findings comply with the predominant wind directions and air velocity in Famagusta, presented in Section 3.2. A 10% WFR needs to be fully opened to provide all the occupancy hours inside category I, while for a 25% WFR, any window orientation having an opening ratio ranging between 25% to fully opened widows can ensure category I of the CO_2 concentration for the 2088 occupancy hours. Cross-ventilation scenarios are more efficient in terms of allowing a greater amount of airflow to pass through openings. Cross-flow by a window combination of the south- and east-facing windows is the most effective case. Conversely, the north- and west-oriented windows offer the least effective cross-ventilation scenario.

Table 16. The most and least effective window orientations for providing a maximum number of acceptable hours based on the CO_2 concentration category I (BS EN 15251:2007).

Ventilation Strategy	Window Size (WFR)	Effective Openings *	Window Openings (%) and Best/Worst Orientations					
			10%		25%		50%, 75%, 100%	
			Best	Worst	Best	Worst	Best	Worst
Single-side	10%	None	West, East	South	West	South	West, East	South
	25%	All openings			All occupancy hours appear in category I			
Cross-flow	10%	None	South + East	North + West	All occupancy hours appear in category I			
	50%	All openings	All occupancy hours appear in category I					

* Comparing different window sizes for the same ventilation strategy.

Table 17 outlines the most and least effective window orientations, in terms of providing a maximum number of acceptable hours according to the European adaptive comfort categories, against different ventilation strategies, window sizes, and opening ratios. In the case of small windows, the least amount of airflow cannot overcome the overheating problem caused by internal gains and direct solar radiation. Therefore, northern windows (in the case of single-side ventilation) as well as north- and west/east-facing windows (in the case of cross-ventilation) provide more acceptable hours of the European adaptive comfort categories due to their receiving a lesser amount of solar radiation. The southern windows (in the case of single-side ventilation) as well as a combination of the south- and east-facing windows (in the case of cross-ventilation) present less effective scenarios. Nevertheless, larger window sizes and opening ratios allow a greater amount of fresh air, from the predominant wind directions of the study location, to enter and cool the spaces; thus, southern windows, as well as south- and east/west-facing windows, turn out to be more effective window orientations.

In general, northwest zones performed better compared to southeast zones on all the floors. Referring to a previous study [5], one interpretation for this situation might be the lesser amount of solar radiation received by those zones due to unshaded windows and inappropriate window material. When a zone has a north-facing window, a greater number of comfortable hours can be achieved. West-oriented windows come in at the second position, followed by the east- and south-oriented windows, respectively. Owing to the fact that unshaded south windows can result in the overheating of internal spaces, one can perceive that in the cases of closed and 10% opened windows, the south-facing windows produce thermally uncomfortable indoor environments. In these cases, the amount of airflow from natural ventilation cannot confront the elevated temperature from external and internal gains. Therefore, the zones with south-oriented windows can have minimal comfortable hours based on adaptive comfort categories.

Table 17. The most and least effective window orientations for providing a maximum number of acceptable hours based on the European adaptive comfort (BS EN 15251:2007).

Ventilation Strategy	Window Size (WFR)	Effective Openings *	Window Openings (%) and Best/Worst Orientations					
			10%		25%		50%, 75%, 100%	
			Best	Worst	Best	Worst	Best	Worst
Single-side	10%	All openings	North	South	North, West	East	South	North
	25%	None	North	South	West	South		
Cross-flow	10%	10%, 25%	North + West	South + East	South + East	North + West	South + East	North + West
	50%	50%, 75%, 100%	North + East		South + West		South + West	

* Comparing different window sizes for the same ventilation strategy.

However, it was observed that three-quarter and full window openings result in a less effective window and natural ventilation relationship in terms of thermal comfort performance compared to quarter and half window openings. This is because larger opening portions can increase the risk of overheating and overcooling on the indoor environment due to the extreme outdoor conditions in both summer and winter periods. Furthermore, larger window areas and opening ratios allow a greater amount of airflow from natural ventilation, while this does not guarantee improved indoor thermal conditions. A larger window area contributes to more heat gain and loss if a suitable window material is not selected or the window area is not protected from direct sun radiation. In contrast to the 10% WFR case, increasing the opened portion for south- and east-facing windows offer more hours in category I and II on each floor. The other window orientations reduce their efficiency with a larger window opening area regardless of the floor location. In the case of a fully glazed external wall, cross-ventilation improves indoor thermal comfort when increasing window-opening ratios.

5.2. Concluding Remarks and Recommendations

In the Mediterranean climate, window-based natural ventilation has a significant potential to improve indoor environmental conditions in free-running period. Therefore, the effectiveness of natural ventilation is extremely associated with early window design and post-occupancy user behaviour. In naturally ventilated buildings, indoor air quality and thermal comfort have a close correlation with each other, thus an "indoor air quality–thermal comfort" dilemma exists. This study examined the relationship between window design and natural ventilation performance in the Mediterranean office buildings in terms of the level of CO_2 concentration and thermal comfort condition. The study applied an experimental method of computational modelling and simulation utilising TAS Engineering software to perform dynamic thermal simulations. The building was designed as a three-storey office building with four thermal zones on each floor, while different window sizes, orientations, and opening scenarios were studied for both single-side and cross-ventilation strategies. Carbon dioxide concentration categories and the adaptive comfort model were determined and assessed based on the BS EN 15251:2007 standard. The study was limited to a three-storey office building, a floor layout with a 1:1 aspect ratio, common materials in envelope construction of the study location, unshaded windows (neither from external nor from internal sides), and a high-occupancy office. Therefore, it presents the following concluding remarks:

- Closed windows for any window size, orientation and location cannot provide any office working hours that the CO_2 concentration appears under category I and II according to the BS EN 15251: 2007 standard. In addition, the CO_2 level exceeds the recommended threshold (1000 ppm); it also reaches 2000 ppm, for which occupants may suffer from sick building syndrome (SBS).
- In the free-running period, a window opening is the main method of ventilation and cooling, thus occupants use windows as well as other physiological adaptation mechanisms to maintain indoor air and thermal conditions. Therefore, closing windows is not acceptable, neither for indoor air nor for thermal comfort conditions, even in the winter months.
- Natural ventilation performance depends on the direction of the wind, air velocity, and the turbulence characteristics of the wind.
- From an architectural point of view, window design, including various parameters, highly effects natural ventilation performance. Thus, architects should study and understand the relationship between window design and natural ventilation in a particular climatic condition, to help them make informed decisions in the early design stage.
- Cross-ventilation scenarios are more efficient in terms of allowing a greater amount of airflow to pass through openings. Cross-flow by a window combination of the south- and east-facing windows is the most effective case. Conversely, the north- and west-oriented windows offer the least effective cross-ventilation scenario.
- Despite the existence of a cross-ventilation strategy, the sun's harmful rays could reduce the potential of this effective passive strategy. It was observed that larger window sizes and opening ratios could decrease the effectiveness of window and natural ventilation due to the extreme outdoor weather conditions in both the summer and winter months.
- Overall, the results of unshaded windows of this study indicate that single-sided ventilation through a small window size (i.e., 10% WFR) with half to fully opened area can be more effective than larger window sizes of the same ventilation strategy, and even more effective than cross-ventilation of various window designs in adjacent walls.
- Floor location has its effect on the window and natural ventilation performance in a way that the windows of the higher floor zones are more effective than those in the lower floors do.
- Natural ventilation performance decreases in the first-floor zones, showing higher carbon dioxide levels, namely for the south-facing window in the summer and north-facing window in the winter.
- Natural ventilation performance shows less efficient in terms of diluting CO_2 contaminant in the cool period compared to the warm period.

- Unshaded windows, even with the most effective design and ventilation strategy, can only provide 50% to 60% of the office occupancy time as thermally acceptable for adaptive thermal comfort.
- To adopt passive design strategies effectively in the Mediterranean climatic, it is important to consider every building envelope element, such as the optimal window design attributes, window-to-floor area, window type, appropriate glazing materials, window orientation, and the required shading ratios to improve indoor thermal comfort and reduce CO_2 levels. More studies are required to address conflicting performance criteria simultaneously in naturally ventilated office buildings.
- A performance-based window design model can guide architects toward making knowledge-based and informed-decisions in the early architectural design stages.

Author Contributions: H.K.A. and H.Z.A. conceived and designed the concept and outline for the article; H.K.A. performed the experiments, simulations, and wrote the manuscript; H.Z.A. supervised, provided sources, comments, and suggestions for the paper. All authors have read and agreed to the published version of the manuscript.

Funding: This research received no external funding.

Conflicts of Interest: The authors declare no conflict of interest.

References

1. Lai, H.K.; Kendall, M.; Ferrier, H.; Lindup, I.; Alm, S.; Hänninen, O.; Jantunen, M.; Mathys, P.; Colvile, R.; Ashmore, M.R.; et al. Personal Exposures and Microenvironment Concentrations of $PM_{2.5}$, VOC, NO_2 and CO in Oxford, UK. *Atmos. Environ.* **2004**, *38*, 6399–6410. [CrossRef]
2. DeDear, R.J.; Brager, G.S. Thermal Comfort in Naturally Ventilated Buildings: Revisions to ASHRAE Standard 55. *Energy Build.* **2002**, *34*, 549–561. [CrossRef]
3. Fang, L.; Wargocki, P.; Witterseh, T.; Clausen, G.; Fanger, P.O. Field Study on the Impact of Temperature, Humidity and Ventilation on Perceived Air Quality. In Proceedings of the Indoor Air Quality and Climate, Edinburgh, Scotland, 8–13 August 1999; pp. 107–112.
4. Larsen, T.S.; Heiselberg, P. Single-sided natural ventilation driven by wind pressure and temperature difference. *Energy Build.* **2008**, *40*, 1031–1040. [CrossRef]
5. Abdullah, H.K.; Alibaba, H.Z. Towards Nearly Zero-Energy Buildings: The Potential of Photovoltaic-Integrated Shading Devices to Achieve Autonomous Solar Electricity and Acceptable Thermal Comfort in Naturally ventilated Office Spaces. In Proceedings of the 16th International Conference on Clean Energy, Famagusta, North Cyprus, 9–11 May 2018; pp. 1–11.
6. Mora-pérez, M.; Guillen-guillamón, I.; López-patiño, G.; López-jiménez, P.A. Natural Ventilation Building Design Approach in Mediterranean Regions—A Case Study at the Valencian Coastal Regional Scale (Spain). *Sustainability* **2016**, *8*, 855. [CrossRef]
7. Wong, N.H.; Huang, B. Comparative Study of the Indoor Air Quality of Naturally Ventilated and Air-Conditioned Bedrooms of Residential Buildings in Singapore. *Build. Environ.* **2004**, *39*, 1115–1123. [CrossRef]
8. Omrani, S.; Garcia-Hansen, V.; Capra, B.R.; Drogemuller, R. On the Effect of Provision of Balconies on Natural Ventilation and Thermal Comfort in High-Rise Residential Buildings. *Build. Environ.* **2017**, *123*, 504–516. [CrossRef]
9. Taylor, P.; Liping, W.; Hien, W.N. Applying Natural Ventilation for Thermal Comfort in Residential Buildings in Singapore. *Archit. Sci. Rev.* **2011**, *50*, 224–233. [CrossRef]
10. Dascalaki, E.G.; Sermpetzoglou, V.G. Energy Performance and Indoor Environmental Quality in Hellenic Schools. *Energy Build.* **2011**, *43*, 718–727. [CrossRef]
11. Sharmin, T.; Gül, M.; Li, X.; Ganev, V.; Nikolaidis, I.; Al-Hussein, M. Monitoring Building Energy Consumption, Thermal Performance, and Indoor Air Quality in a Cold Climate Region. *Sustain. Cities Soc.* **2014**, *13*, 57–68. [CrossRef]
12. Lei, Z.; Liu, C.; Wang, L.; Li, N. Effect of Natural Ventilation on Indoor Air Quality and Thermal Comfort in Dormitory during Winter. *Build. Environ.* **2017**, *125*, 240–247. [CrossRef]

13. Inanici, M.N.; Demirbilek, F.N. Thermal Performance Optimization of Building Aspect Ratio and South Window Size in Five Cities Having Different Climatic Characteristics of Turkey. *Build. Environ.* **2000**, *35*, 41–52. [CrossRef]

14. Baker, N.; Steemers, K. *Energy and Environment in Architecture*; Taylor & Francis: London, UK, 2003.

15. Steadman, P.; Bruhns, H.R.; Holtier, S.; Gakovic, B. A Classification of Built Forms. *Environ. Plan. B Urban Anal. City Sci.* **2000**, *27*, 73–91. [CrossRef]

16. Ratti, C.; Baker, N.; Steemers, K. Energy Consumption and Urban Texture. *Energy Build.* **2005**, *37*, 762–776. [CrossRef]

17. Lin, M.; Pan, Y.; Long, W.; Chen, W. Influence of Building Shape Coefficient on Energy Consumption of Office Buildings in Hot-Summer-and-Cold-Winter Area of China. *Build. Energy Effic.* **2015**, *10*, 728–735.

18. St. Clair, P.; Hyde, R. Towards a New Model for Climate Responsive Design at the University of the Sunshine Coast Chancellery. *J. Green Build.* **2010**, *4*, 3–20. [CrossRef]

19. Chua, K.J.; Chou, S.K. Energy Performance of Residential Buildings in Singapore. *Energy* **2010**, *35*, 667–678. [CrossRef]

20. Mirrahimi, S.; Mohamed, M.F.; Haw, L.C.; Ibrahim, N.L.N.; Yusoff, W.F.M.; Aflaki, A. The Effect of Building Envelope on the Thermal Comfort and Energy Saving for High-Rise Buildings in Hot-Humid Climate. *Renew. Sustain. Energy Rev.* **2016**, *53*, 1508–1519. [CrossRef]

21. Ghafari, F.; Mirrahimi, S.Z.; Heidari, S. Influence of Ceiling Height on Heating Energy Consumption in Educational Building. In Proceedings of the 15th International Conference on Civil and Architecture Engineering, Pattaya, Thailand, 25–26 April 2018; pp. 1–7.

22. Guimarães, R.P.; Carvalho, M.C.R.; Santos, F.A. The Influence of Ceiling Height in Thermal Comfort of Buildings: A Case Study in Belo Horizonte, Brazil. *Int. J. Hous. Sci.* **2013**, *37*, 75–85.

23. Ralegaonkar, R.V.; Gupta, R. Review of Intelligent Building Construction: A Passive Solar Architecture Approach. *Renew. Sustain. Energy Rev.* **2010**, *14*, 2238–2242. [CrossRef]

24. Mickaël, D.; Bruno, B.; Valérie, C.; Murielle, L.; Cécile, P.; Jacques, R.; Severine, K. Indoor Air Quality and Comfort in Seven Newly Built, Energy-Efficient Houses in France. *Build. Environ.* **2014**, *72*, 173–187. [CrossRef]

25. Zhu, C.; Li, N. Study on Indoor Air Quality Evaluation Index Based on Comfort Evaluation Experiment. *Procedia Eng.* **2017**, *205*, 2246–2253. [CrossRef]

26. Wang, J.; Li, J. Bio-Inspired Kinetic Envelopes for Building Energy Efficiency Based on Parametric Design of Building Information Modeling. In Proceedings of the 2010 Asia-Pacific Power Energy Engineering Conference (APPEEC), Chengdu, China, 28–31 March 2010; pp. 1–4.

27. Abdullah, H.K.; Alibaba, H.Z. Retrofits for Energy Efficient Office Buildings: Integration of Optimized Photovoltaics in the Form of Responsive Shading Devices. *Sustainability* **2017**, *9*, 2096. [CrossRef]

28. Alibaba, H.Z.; Ozdeniz, M.B. Energy Performance and Thermal Comfort of Double-Skin and Single-Skin Facades in Warm-Climate Offices. *J. Asian Archit. Build. Eng.* **2016**, *15*, 635–642. [CrossRef]

29. Al-Tamimi, N.A.M.; Fadzil, S.F.S.; Harun, W.M.W. The Effects of Orientation, Ventilation, and Varied WWR on the Thermal Performance of Residential Rooms in the Tropics. *J. Sustain. Dev.* **2011**, *4*, 142–149. [CrossRef]

30. Stabile, L.; Dell'Isola, M.; Russi, A.; Massimo, A.; Buonanno, G. The Effect of Natural Ventilation Strategy on Indoor Air Quality in Schools. *Sci. Total Environ.* **2017**, *595*, 894–902. [CrossRef]

31. Costanzo, V.; Evola, G.; Marletta, L. A Review of Daylighting Strategies in Schools: State of the Art and Expected Future Trends. *Buildings* **2017**, *7*, 41. [CrossRef]

32. International Organization for Standardization. *EN ISO 16814—Building Environment Design—Indoor Air Quality—Methods of Expressing the Quality of Indoor Air for Human Occupancy*; International Organization for Standardization: Geneva, Switzerland, 2008.

33. Fadzil, S.F.S.; Abdullah, A.; Harun, W.M.W. The Impact of Varied Orientation and Wall Window Ratio (WWR) To Daylight Distribution in Residental Rooms. In Proceedings of the International Symposium on Construction in Developing Economies: Commonalities among Diversities (CIBW 107), Penang, Malaysia, 5 Octobar 2009; pp. 478–485.

34. Wagdy, A.; Sherif, A.; Sabry, H.; Arafa, R.; Mashaly, I. Daylighting Simulation for the Configuration of External Sun-Breakers on South Oriented Windows of Hospital Patient Rooms under a Clear Desert Sky. *Sol. Energy* **2017**, *149*, 164–175. [CrossRef]

35. AlAnzi, A.; Seo, D.; Krarti, M. Impact of Building Shape on Thermal Performance of Office Buildings in Kuwait. *Energy Convers. Manag.* **2009**, *50*, 822–828. [CrossRef]

36. Pedrini, A.; Vilar De Carvalho, J.P. Analysis of Daylight Performance in Classrooms in Humid and Hot Climate. In Proceedings of the 30th International PLEA Conference, Ahmedabad, India, 16–18 December 2014; pp. 1–10.

37. Aflaki, A.; Mahyuddin, N.; Al-Cheikh Mahmoud, Z.; Baharum, M.R. A Review on Natural Ventilation Applications through Building Façade Components and Ventilation Openings in Tropical Climates. *Energy Build.* **2015**, *101*, 153–162. [CrossRef]

38. Alibaba, H.Z. Determination of Optimum Window to External Wall Ratio for Offices in a Hot and Humid Climate. *Sustainability* **2016**, *8*, 187. [CrossRef]

39. Alibaba, H.Z. Heat and Air Flow Behavior of Naturally Ventilated Offices in a Mediterranean Climate. *Sustainability* **2018**, *10*, 3284. [CrossRef]

40. Fan, G.; Xie, J.; Liu, J. Indoor Air Quality in a Naturally Ventilated Research Student Office in Chinese Universities during Heating Period. *Procedia Eng.* **2017**, *205*, 1272–1278. [CrossRef]

41. Laska, M.; Dudkiewicz, E. Research of CO_2 Concentration in Naturally Ventilated Lecture Room. In Proceedings of the International Conference on Advances in Energy Systems and Environmental Engineering (ASEE17), Wrocław, Poland, 2–5 July 2017; pp. 1–8. [CrossRef]

42. Krawczyk, D.A.; Rodero, A.; Gładyszewska-Fiedoruk, K.; Gajewski, A. CO_2 concentration in Naturally Ventilated Classrooms Located in Different Climates—Measurements and Simulations. *Energy Build.* **2016**, *129*, 491–498. [CrossRef]

43. Ye, W.; Zhang, X.; Gao, J.; Cao, G.; Zhou, X.; Su, X. Indoor Air Pollutants, Ventilation Rate Determinants and Potential Control Strategies in Chinese Dwellings: A Literature Review. *Sci. Total Environ.* **2017**, *586*, 696–729. [CrossRef]

44. WHO. *Air Quality Guidelines for Europe*, 2nd ed.; World Health Organization Regional Office for Europe, European Series, No. 91; WHO: Copenhagen, Denmark, 2000. [CrossRef]

45. Jones, A.P. Chapter 3 Indoor Air Quality and Health. *Dev. Environ. Sci.* **2002**, *1*, 57–115. [CrossRef]

46. Al Horr, Y.; Arif, M.; Katafygiotou, M.; Mazroei, A.; Kaushik, A.; Elsarrag, E. Impact of Indoor Environmental Quality on Occupant Well-Being and Comfort: A Review of the Literature. *Int. J. Sustain. Built Environ.* **2016**, *5*, 1–11. [CrossRef]

47. Erdman, C.A.; Steiner, K.C.; Apte, M.G. Indoor Carbon Dioxide Concentrations and Sick Building Syndrome Symptoms in the Base Study Revisited: Analyses of the 100 Building Dataset. In Proceedings of the International Conference on Indoor Air, Monterey, CA, USA, 30 June–5 July 2002; Volume III, pp. 443–448.

48. Vasile, V.; Petran, H.; Dima, A.; Petcu, C. Indoor Air Quality—A Key Element of the Energy Performance of the Buildings. *Energy Procedia* **2016**, *96*, 277–284. [CrossRef]

49. Daniel, S.; Rubio-bellido, C.; Pulido-arcas, J.A. Adaptive Comfort Models Applied to Existing Dwellings in Mediterranean Climate Considering Global Warming. *Sustainability* **2018**, *10*, 3507. [CrossRef]

50. Campano, M.Á.; Dom, S.; Fern, J.; Jos, J. Thermal Perception in Mild Climate: Adaptive Thermal Models for Schools. *Sustainability* **2019**, *11*, 3948. [CrossRef]

51. Castaño-Rosa, R.; Rodríguez-Jiménez, C.E.; Rubio-Bellido, C. Adaptive Thermal Comfort Potential in Mediterranean Office Buildings: A Case Study of Torre Sevilla. *Sustainability* **2018**, *10*, 3091. [CrossRef]

52. Salvalai, G.; Pfafferott, J.; Sesana, M.M. Assessing energy and thermal comfort of different low-energy cooling concepts for non-residential buildings. *Energy Convers. Manag.* **2013**, *76*, 332–341. [CrossRef]

53. Engelmann, P.; Kalz, D.; Salvalai, G. Cooling concepts for non-residential buildings: A comparison of cooling concepts in different climate zones. *Energy Build.* **2014**, *82*, 447–456. [CrossRef]

54. Croitoru, C.; Nastase, I.; Crutescu, R.; Badescu, V. Thermal comfort in a Romanian passive house. Preliminary results. *Energy Procedia* **2018**, *85*, 575–583. [CrossRef]

55. EDSL. *TAS Engineering*; Environmental Design Solutions Ltd. (EDSL): Stony Stratford, UK, 2019.

56. Peel, M.C.; Finlayson, B.L.; McMahon, T.A. Updated World Map of the Köppen-Geiger Climate Classification. *Hydrol. Earth Syst. Sci.* **2007**, *11*, 1633–1644. [CrossRef]

57. Kottek, M.; Grieser, J.; Beck, C.; Rudolf, B.; Rubel, F. World Map of the Köppen-Geiger Climate Classification Updated. *Meteorol. Zeitschrift* **2006**, *15*, 259–263. [CrossRef]

58. Union of The Chambers of Cyprus Turkish Engineers and Architects (KTMMOB). Chapter 96 Regulation: The Streets and Buildings Regulation. Available online: http://www.mimarlarodasi.org/tr/mevzuat/fasil-96-duzenleme/ (accessed on 21 May 2019).
59. CIBSE. *Guide A: Environmental Design*; The Chartered Institution of Building Services Engineers (CIBSE): London, UK, 2015.
60. ASHRAE. *ANSI/ASHRAE Standard 55—Thermal Environmental Conditions for Human Occupancy*; American National Standard Institute (ANSI) and American Society of Heating, Refrigerating, and Air-Conditioning Engineers (ASHRAE): Atlanta, GA, USA, 2017.
61. ASHRAE. *ASHRAE Handbook: Fundamentals*; American Society of Heating, Refrigerating, and Air-Conditioning Engineers: Atlanta, GA, USA, 2013.
62. ASHRAE. *ANSI/ASHRAE Standard 62.1—Ventilation for Acceptable Indoor Air Quality*; American National Standard Institute (ANSI) and American Society of Heating, Refrigerating, and Air-Conditioning Engineers (ASHRAE): Atlanta, GA, USA, 2019.
63. Voss, J. *Revisiting Office Space Standards.*; Haworth, Inc.: Grand Rapids, MI, USA, 2000.
64. EDSL. *TAS Theory Manual*; Environmental Design Solutions Ltd. (EDSL): Stony Stratford, UK, 2019; Available online: http://www.edsl.net/main/Support/Documentation.aspx (accessed on 13 December 2019).
65. British Standards Institution (BSI). *BS EN 15251—Indoor Environmental Input Parameters for Design and Assessment of Energy Performance of Buildings Addressing Indoor Air Quality, Thermal Environment, Lighting and Acoustics*; British Standards Institution (BSI): London, UK, 2007.
66. British Standards Institution (BSI). *BS EN 13779—Ventilation for Non-Residential Buildings—Performance Requirements for Ventilation and Room-Conditioning Systems*; British Standards Institution (BSI): London, UK, 2007.
67. Fanger, P.O. *Thermal Comfort*; Danish Technical Press: Copenhagen, Denmark, 1970.
68. Nguyen, A.T.; Singh, M.K.; Reiter, S. An Adaptive Thermal Comfort Model for Hot Humid South-East Asia. *Build. Environ.* **2012**, *56*, 291–300. [CrossRef]
69. DeDear, R.J.; Brager, G.S. Developing an Adaptive Model of Thermal Comfort and Preference. *ASHRAE Trans.* **1998**, *104*, 145–167.
70. Nicol, F.; Roaf, S. Pioneering: New Indoor Temperature Standards the Pakistan Project. *Energy Build.* **1996**, *23*, 169–174. [CrossRef]
71. Humphreys, M.A.; Nicol, J.F.; Raja, I.A. Field Studies of Indoor Thermal Comfort and the Progress of the Adaptive Approach. *Adv. Build. Energy Res.* **2007**, *1*, 55–88. [CrossRef]
72. Nicol, F.; Humphreys, M. Derivation of the Adaptive Equations for Thermal Comfort in Free-Running Buildings in European Standard EN15251. *Build. Environ.* **2010**, *45*, 11–17. [CrossRef]

 *sustainability*

Article

Reduction Strategies for Greenhouse Gas Emissions from High-Speed Railway Station Buildings in a Cold Climate Zone of China

Nan Wang [1], Daniel Satola [2], Aoife Houlihan Wiberg [3], Conghong Liu [1],* and Arild Gustavsen [2]

[1]   School of Architecture, Tianjin University, No. 92 Weijin Street, Nankai District, Tianjin 300072, China; nancywang@tju.edu.cn
[2]   Research Centre of Zero Emission Neighbourhoods in Smart Cities (FME-ZEN), Department of Architecture and Technology, Norwegian University of Science and Technology, 7491 Trondheim, Norway; daniel.satola@ntnu.no (D.S.); arild.gustavsen@ntnu.no (A.G.)
[3]   Belfast School of Architecture and the Built Environment, Faculty of Computing, Engineering and the Built Environment, Ulster University, Belfast BT15 1ED, UK; a.wiberg@ulster.ac.uk
*   Correspondence: conghong_liu@163.com

Received: 21 January 2020; Accepted: 18 February 2020; Published: 25 February 2020

Abstract: Implementing China's emission reduction regulations requires a design approach that integrates specific architectural and functional properties of railway stations with low greenhouse gas (GHG) emission. This article analyzes life cycle GHG emissions related to materials production, replacement and operational energy use to identify design drivers and reduction strategies implemented in high-speed railway station (HSRS) buildings. A typical middle-sized HSRS building in a cold climate zone in China is studied. A detailed methodology was proposed for the development and assessment of emission reduction strategies through life cycle assessment (LCA), combined with a building information model (BIM). The results reveal that operational emissions contribute the most to total GHG emissions, constituting approximately 81% while embodied material emissions constitute 19%, with 94 $kgCO_{2eq}/m^2 \cdot a$ and 22 $kgCO_{2eq}/m^2 \cdot a$ respectively. Optimizing space can reduce operational GHG emissions and service life extension of insulation materials contributes to a 15% reduction in embodied GHG emissions. In all three scenarios, the reduction potentials of space, envelope, and material type optimization were 28.2%, 13.1%, and 3.5% and that measures for reduced life cycle emissions should focus on space in the early stage of building design. This study addresses the research gap by investigating the life cycle GHG emissions from HSRS buildings and reduction strategies to help influence the design decisions of similar projects and large space public buildings which are critical for emission reduction on a larger scale.

Keywords: GHG emissions; life cycle assessment; large space building; high-speed railway station

1. Introduction

Globally, the construction and operation of buildings account for 36% of the total final energy use and nearly 40% of greenhouse gas (GHG) emissions represented by carbon dioxide (CO_2) [1]. China's carbon emissions will peak by 2030 and, thus, need to be reduced dramatically as pledged in the Intended Nationally Determined Contribution (INDC), signed as part of the Paris Climate Agreement [2]. As the largest carbon emitter in the world, China is facing severe challenges in saving energy and reducing GHG emissions of built environments, which are gaining in importance as it moves to bring the building sector under control. In the 13th Five-Year Plan (2016–2020) implemented by the State Council, aiming to save energy and cut emissions, public buildings are emphasized [3], given that they have the largest share in the total consumption of operational energy (excluding district heating) among the different building categories. The most important reason is that the proportion of large-scale or large space public buildings in the building sector—for instance, transportation buildings—has been rising rapidly [4]. Large space buildings refer to those that have great floor height, large enclosed space volumes, and diverse ventilation systems [5]. Energy consumption of transportation buildings per square metre is two to three times that of conventional public buildings such as offices and schools, indicating their greater impact on energy and environment [6] (see Figure 1). As a typical large space public building, a high-speed railway station (HSRS) building is attracting increasing attention. The operating mileage of high-speed railway in China currently ranks first in the world under the 'eight vertical and eight horizontals' railway network plan [7]. These often run on newly created lines and therefore require improved modern stations on new sites. HSRS, known as the gateway to the city, will gradually become the most common type of railway stations, and an increasing amount of constructions will likely promote an upward trend in energy and resource demand in the immediate future. Due to increasingly stringent energy requirements and improved energy efficiency, the embodied GHG emissions from building materials are gaining significance [8–11]. As a major consumer of societal energy and national resources, it is necessary to conduct evaluative research on HSRS buildings in the scope of embodied GHG emissions.

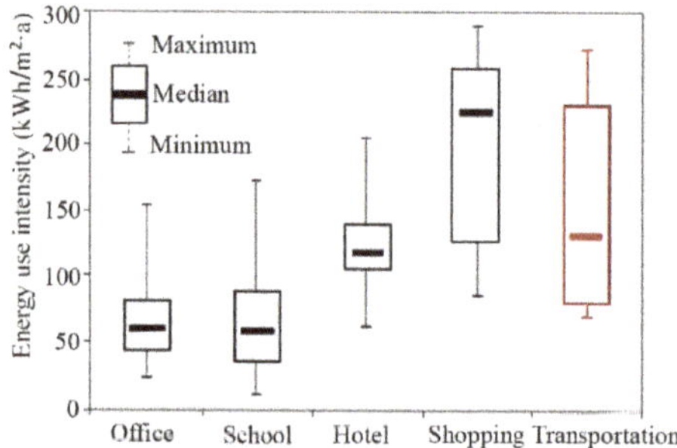

Figure 1. Energy use comparison among five types of public buildings [6].

Several studies have focused on the energy consumption and carbon emissions from multiple building typologies internationally, including residential, office, school, and hotel buildings in China [8,9], Norway [10,11], Spain [12] and the United Kingdom [13]. More attention should be paid to the environmental performance of HSRS buildings, which also needs to be drastically improved. A shift toward low-emission buildings of railway stations is laying down solid tracks for the sustainable building sector. The concepts of 'zero emission station', 'zero energy station', 'green station', and 'eco station' have emerged in order to address the global warming challenges from railway stations: for instance, the European Commission supports sustainable station constructions with the 'Sus Station' project and has achieved a new sustainable, low-carbon generation [14]. The green railway stations rating program of the Indian Green Building Council aims to reduce the adverse environmental impacts due to station operation and maintenance [15]. Despite all of this, earlier studies on railway station buildings mainly focused on energy efficiency [16–22], indoor thermal comfort [23,24], and ventilation and air quality evaluation [25,26]. Some research discussed the environmental impacts of the railway transport sector, in which railway stations are usually considered as a part of the whole infrastructure. In rail system studies, Håkan et al. [27] presented the energy use and GHG emissions results from example railway stations in construction, maintenance and operation by a life cycle approach. Zhang et al. [28] calculated high-speed transport energy consumption based on the construction investment integrating elements of life cycle analysis. However, work on the direct perspective of the emission data, as well as design measures in detail, is rather limited. Further study is needed to consider life cycle GHG emissions from HSRS buildings, which is rarely studied as the main research object.

Life cycle assessment (LCA) is a well established approach used for the environmental assessment of buildings and provides the necessary information for reducing the environmental impacts systematically and comprehensively [29]. It would be only giving a limited perspective of the environmental impact of a given building to judge the GHG emissions only during its operating phases. LCA makes it possible to evaluate the total GHG emissions and help to determine the ratio between embodied and operational emissions. Furthermore, LCA also aids designers in selecting optimal design solutions, by evaluating different emission reduction measures [30].

The complete life cycle of a building measures the cradle-to-grave impacts from four main stages [31]: the product stage (A1–A3), construction process stage (A4–A5), use stage (B1–B7), and end-of-life stage (C1–C4). The optional stage of benefits and loads (D) is defined to document potential emission compensation of processing or reusing materials after end-of-life. In accordance with international LCA standards [32,33], the LCA methodology is used to quantify the life cycle GHG emissions following four main steps: goal and scope, life cycle inventory (LCI), environmental impacts assessment, and results interpretation. Different studies have carried out the life cycle GHG emission calculations with varying objectives, system boundaries, data sources, and levels of detail, following the LCA methodology [11,34,35]. Some studies have briefly evaluated the life cycle CO_2 emissions of an HSRS building for energy efficient and carbon reduction analysis [21,22]. However, these previous studies have not analyzed the purpose of evaluating life cycle GHG emissions to extract important design drivers and further understand how to reduce GHG emissions for HSRS buildings. In general, research on life cycle GHG emissions from HSRS buildings is representative, and provides a scientific basis for other large space public buildings.

The need to study the profile of environmental performance of HSRS buildings has led the authors to explore adaptive carbon reduction strategies on low GHG emission design. A typical medium-sized HSRS building in Tianjin, China was selected for the present study as a subject on which to conduct a detailed analysis. This work begins with introducing HSRS in China and outlining significant carbon mitigation measures in general buildings collected from literature, which provide background information about the following analysis. Section 3 presents the methodology used for evaluating and discussing the case study, taking into consideration the life cycle GHG emissions generated from materials and operational energy use. Section 4 describes the case building. The embodied, operational, and total GHG emissions are calculated and results are presented in Section 5 to extract design contributors and deduce reduction-oriented aspects. These findings are further discussed and a stepwise analysis from single to multiple strategies is proposed. The updated results of GHG emissions related to the case with different strategies regarding building space, envelope, and materials are recalculated, and the reduction proportion between origin and updated results of GHG emissions are investigated to extract reduction potentials by comparison. Final remarks are drawn in the conclusions.

2. Background and Related Studies

2.1. HSRS in China

2.1.1. Driving Policies and Regulations

For sustainable development and low carbon transformation, all levels of government (local, province, and even transportation agencies) have adopted effective policies to strengthen response, and taken actions to combat climate change, as illustrated later. A special plan for building energy conservation in China promulgated energy saving renovations at airports, piers, and railway stations according to China's Energy Policy 2012 [36]. The 13th Five-Year Plan (2016–2020) continues to upgrade public buildings to high performance green buildings. Large railway stations are required to optimize design to realize gains in energy efficiency. The China Railway Corporation (CRC) pushed forward energy conservation in railway stations [7] and was required to stimulate a green and low carbon transportation plan in accordance with the 13th Five-Year Plan of Railway Development [37]. Newly built railway stations must meet the requirements of the design standard for energy efficiency of public buildings (GB 50189-2015) [38], which indicate that energy consumption of new buildings should be lower than 75% of that of comparable buildings in the 1980s. The standards for drainage systems, electrical systems, and renewable energy have also been raised [29]. Despite stricter design requirements and improved energy saving standards, there are few systematic design principles, strategies, and applications for reducing the life cycle GHG emissions of HSRS buildings. Under most circumstances, they only rely upon conventional building solutions and experiences of generic buildings and do not consider specific characteristics.

2.1.2. Characteristics of HSRS Buildings and Climate Zones in China

Railway stations are classified into super large, large, medium, and small sizes by construction scale, based on the maximum assembling passenger number or dispatched passenger number during rush hour [39]. Medium-sized railway stations are significantly important for study because they have the largest share of the railway station stock in China. A railway station is a complex example of civil engineering, as it includes buildings, platforms, and other infrastructures. In this study, a railway station building refers to the main building in the station, a multi-functional complex that provides facilities such as a waiting room, ticket office, shopping, restaurants, variety stores, and other amenities for passengers. Compared with general public buildings such as governmental offices, domestic HSRS buildings have certain characteristics that are similar to the renewed modern railway stations. Typically, massive, reinforced concrete structures are used with large glass areas and structural steel (see Figure 2). The waiting room/hall is designed as a large enclosed space volume with high floor height making the building a city landmark. Many new stations follow design principles similar to airports, with constant daily operation [20]. For the convenience of management and use, the station building is usually designed along the railway line, so its main orientation is not consistent with the best orientation in most cases.

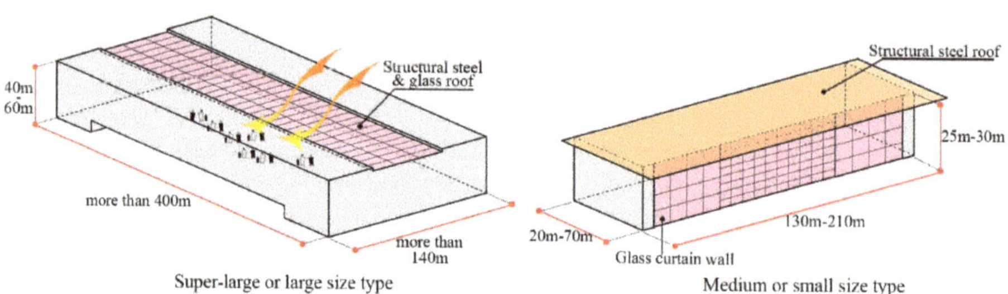

Figure 2. Schematic diagram of HSRS building in shape and scale [40].

The local climate conditions to which the building is exposed play a significant role in selecting effective climate responsive strategies during the design process. As defined by the National Uniform Standard for Design of Civil Buildings (GB 50352-2019), China is divided into five main climate zones: severe cold, cold, hot summer and cold winter, hot summer and warm winter, and mild. In northern China, operating energy has a relatively comprehensive configuration in which heating accounts for a larger proportion compared to that in southern regions because of climate condition differences. Severe cold and cold zones in the north region present saving potentials for operational energy emissions on cooling as well as heating. One of the largest cities in the cold zone, Tianjin, has a typical temperate monsoon climate with distinct seasons, which varies greatly in summer (hot and rainy) and winter (cold and dry). The thermal design of the building envelope in Tianjin requires the extensive use of strategies for insulation, not only for the cold winter weather but also for summer heat protection. In this research, a medium-sized building from Tianjin South Railway Station was selected as the case to be studied (henceforth Case-TJS). The cold zone and site of the case building are depicted in Figure 3.

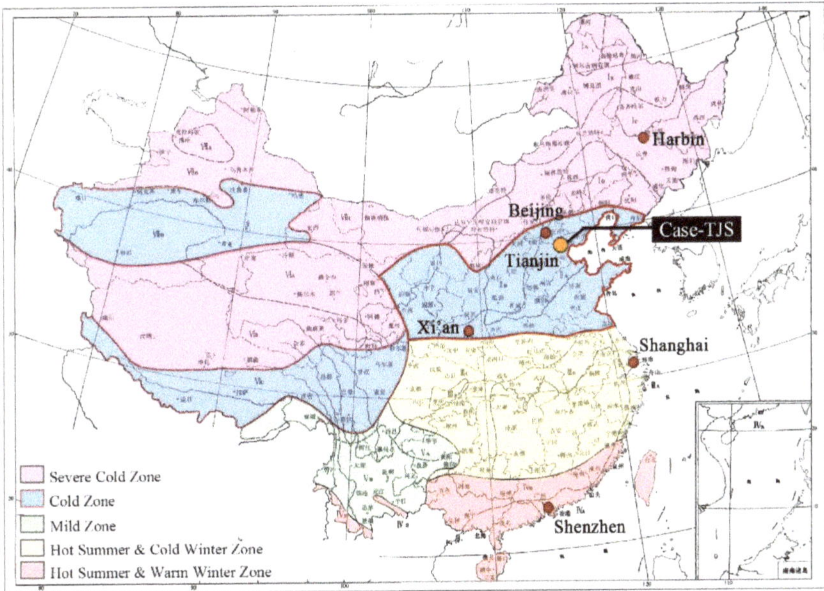

Figure 3. Climatic regionalization map for architecture in China and site of the case study building. (Red spot: representative cities in China, orange spot: Tianjin.).

2.1.3. Field Investigation and Questionnaire

To explore the possible influence of design factors on energy consumption, the research team conducted a field investigation of 10 HSRS buildings along the Beijing–Shanghai high-speed railway. The buildings vary in their capacity, GFA, width, depth, and height. A brief comparison is provided in Table 1. It was found that station buildings face the same architectural and technical design problems and have great energy saving potential in terms of form scale, space utilization, function layout, building details, and operating frequency:

- For medium-sized HSRS buildings, the width is usually from 145 m to 206 m, as well as depth being from 20 to 44 m. The average roof height is approximately 26 m, which is larger in buildings of a super large or large size. There are overcrowded passengers in station (10) but few in station (2), and the capacity design is shown to be not consistent with the scale.
- The height above the enclosed waiting room cannot be used efficiently, since they consume energy for operation, which reflects low utilization of vertical space.
- No thermal partitions are set up in waiting rooms. The HVAC system works for a centralized large room and can't be controlled in a modular way according to the pedestrian volume.
- The glass curtain wall makes the building transparent; however, the shading devices seem not to be enough in summer seasons.

A questionnaire survey was also conducted among 485 passengers on travel behaviour, in which 72% stated that they prefer to arrive at the waiting room less than half an hour before a train's departure instead of waiting for a long time. In fact, high-speed trains depart every five to ten minutes in most stations, implying that ticket gates have to be opened frequently to check tickets. Passengers do not have to stay in the enclosed waiting room for a long time. The waiting room area, defined originally by regulations governing conventional railway stations, encourages unnecessary energy consumption in HSRS.

Table 1. Physical and spatial characteristics of investigated HSRS buildings in the cold zone.

Code *	HSRS	Capacity (Person)	GFA (m^2)	Space (m)		
				Width	Depth	Height
(1)	Beijing South railway station	10,500	252,000	350	195	40
(2)	Tianjin West railway station	5000	104,000	400	145	57
(3)	Jinan West railway station	4000	100,000	192	107	44
(4)	Langfang railway station	1000	9889	170	36	20
(5)	Tianjin South railway station	1000	8669	145	20	30
(6)	Cangzhou West railway station	1200	10,213	168	34	22
(7)	Dezhou East railway station	2000	19,810	206	34	30
(8)	Qufu East railway station	1500	9996	206	35	28
(9)	Zaozhuang railway station	1000	9965	185	37	21
(10)	Xuzhou East railway station	2500	14,984	164	44	28

* denotes that stations (1)–(3) are super large and large sizes, stations (4)–(10) are medium-sized.

2.2. GHG Emission Reduction Strategies for Buildings

Research concentrating on railway station buildings, especially the high speed type, is relatively limited; some of the existing literature takes into account reduction measures for energy consumption that are directly correlated to operational emissions, such as: the relationship between energy consumption and passenger flow density [16]; energy efficiency measures on performance parameters of a building envelope, area ratio of skylight and sun shading mode [17]; energy saving approaches that place greater emphasis on the reduction of air infiltration [18]; triadic relation among lighting comfort level and lighting energy consumption [19]; intelligent control strategies as feasible energy saving solutions [20]; and energy efficient analysis on construction scale, space design, function layout, and operation mode [21,22]. Energy potential in conventional stations or HSRS has been analyzed in different domestic regions. These studies mainly focus on reduction measures in terms of building envelope, while few are concerned with the influence of building form in terms of space design.

In recent years, there has been growing interest in not only the evaluation of embodied emissions, but also the effect of innovative and alternative solutions to reduce total embodied environmental impacts. However, there has been rarcly research on railway station buildings focusing on embodied emissions, which can be understood by analyzing other building categories. For instance, Luo et al. [8] analyzed and calculated the CO_2 emissions of 78 office buildings in the construction materialization stage, obtaining results of 327kgCO_2/m^2 on average (building lifetime is considered to be 50 years). A prediction formula using steel reinforcement, concrete, and wall materials as independent variables was found to predict CO_2 emissions. Existing research displayed high carbon mitigation potentials when taking into consideration the substitution of building materials/components during the early stage design of buildings [41]. Zhang et al. [9] demonstrated that residential and office buildings with a reinforced concrete block masonry structure could reduce carbon emissions by 6–18% compared with either a reinforced or brick concrete structure. Substantial reductions of embodied energy (or carbon) for the use of traditional and locally available building materials were shown, compared to using conventional materials/components in some studies [42,43]. Other studies [44,45] revealed that the use of recycled and reused materials can significantly reduce the embodied energy and carbon emissions in buildings. A study by Maddalena et al. [46] indicated that implementation of materials/components with new and innovative technologies helps reduce the embodied carbon of buildings, such as in the form of "green" types of cements and high performance concrete.

3. Methodology

3.1. Life Cycle GHG Emissions Calculation

3.1.1. Goal and Scope

The goal of this analysis is to evaluate, quantify, and present an overview of life cycle GHG emissions of the HSRS building as a basis for investigating emission reduction strategies. Figure 4 shows the life cycle phases and system boundary for the case study. The system boundary is founded on the modular life cycle system boundaries as in EN15978: 2011 [31] and defined in accordance with the scope of OM: O (operational energy use), which refers to life cycle module B6, the direct carbon emissions resulting from fossil fuels and electricity during the operation process; M (materials) corresponds to life cycle modules A1–A3 and B4, for the production and replacement of building materials, the current scope of embodied GHG emission calculation. Embodied GHG emissions are limited from raw material supply to manufacturing of the main products and materials needed, which include material inputs to the gate. The replacement of new materials over the building's lifetime has also been included. A building reference study period is set to 50 years in terms of the design lifetime (50–100 years) of HSRS based on the standard [47]. The estimated service lifespans of different building materials and components are mainly based on average values from product suppliers and research literature [48,49].

The functional unit is expressed as per 1 m^2 of unit gross floor area (GFA) over the building lifetime, to provide a reference for relevant inputs and outputs, so as to ensure the comparability of calculation results. The global warming potential (GWP) is weighted by carbon dioxide equivalents (CO$_{2eq}$) with the Intergovernmental Panel on Climate Change's 100-year horizon method. Thus, life cycle GHG emissions (noted as LCCO$_{2eq}$ below) (kg CO$_{2eq}$/m^2·a), GHG emissions per unit construction area and one year of building's service life, can be taken as the functional unit of LCCO$_{2eq}$ of an HSRS building. All the calculations follow the principles of an environmental assessment through life cycle analysis. The calculation model is as follows.

$$LCCO_{2eq} = LCCO_{2eq}(A1\text{-}A3) + LCCO_{2eq}(B4) + LCCO_{2eq}(B6) \tag{1}$$

Here, LCCO$_{2eq}$(x) refers to the GHG emissions generated from specific (x) life cycle module (A1–A3, B4 or B6). LCCO$_{2eq}$ represents the total GHG emissions of a function unit in the life cycle. In the complete life cycle of an HSRS building, A1–A3, B4, and B6 are together responsible for more than 90% of total emissions in the case studies [21]. The early design stage has the largest impact on the life cycle stages, while controlling transportation/construction, and the end-of-life stage is difficult [50]. Additionally, energy performance can be improved by architects and engineers by integrating adaptive strategies in the early design stage [51]. Thus, the construction process stage (A4–A5) and end-of-life stage (C1–C4) have been excluded in this study.

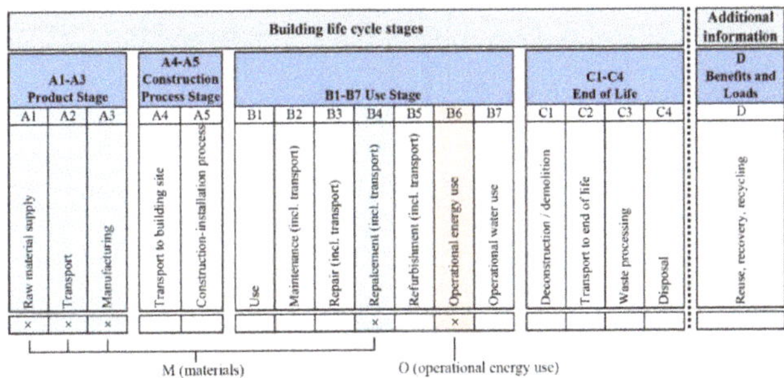

Figure 4. Life cycle phases of a building and system boundary for the case study [31].

3.1.2. Data Sources

Building materials referred to in this study are mainly categorized into four parts: wall materials, concrete, steel, and mortar. When the case study was carried out, there was an unknown environmental product declaration (EPD) providing transparent and consistent information about specific CO$_{2eq}$ emission data from the construction material supplier or manufacturers in China. Generic CO$_{2eq}$ emission data of construction materials from the Chinese national standard for building carbon emission calculation [52] are used, as shown in Table 2. The CO_{2eq} emission data of construction materials in terms of insulation, window, and concrete block are taken from a database of related research [48,49] when the standard is lacking. Electricity is the only form of energy delivered to the case building systems. The annual electric GHG emission factor plays an important role in operational emission calculation. Previous research has updated the CO_{2eq} factors for Chinese electricity grids in seven regions to determine the organization's and product's carbon footprint with necessary modifications [53]. The present work considers a specific factor of 1.15 $kgCO_{2eq}/kWh$ [53] to calculate the emissions from the electricity, and it is assumed to remain constant during the building's lifetime. The GHG emissions related to energy use can be highly sensitive to the electricity emission factor. The current electricity consumed is from the north region power grid, national public network. The primary energy resources involved come from coal, hydroelectric and wind. Coal-fired electricity generation constitutes approximately 99% of power generation, leading to a high value of CO_{2eq} factor.

Table 2. CO_{2eq} emission data of partial construction materials in the standard [52].

Material Type	C30 Reinforced Concrete	Steel	Cement	Sand	Stone
Unit	$kgCO_{2eq}/m^3$	$kgCO_{2eq}/t$	$kgCO_{2eq}/t$	$kgCO_{2eq}/t$	$kgCO_{2eq}/t$
CO_{2eq} Emission Data	295	2340	735	2.51	5.08

3.1.3. Material Inventories and Levels of Detail

Load-bearing structures and building envelope partitions are the main sections in the entire building formation stage [8]. Hence, in this study materials inventory is simplified and excludes inventory of building services, equipment, and internal finishes. The life cycle material inventories for the case study are structured according to the building elements classified in the LCA tool, such as groundwork and foundations, load bearing structure, outer walls, inner walls, floor structure, and outer roof. In addition, the construction elements of railway stations outside surroundings such as railway platforms, platform canopies and loading platforms have not been included in the models due to the focus on developing solutions at the individual building level. The data include the most up-to-date building details; material properties and related information were collected from building constructions in order to build a model to obtain the material quantities data. A quota calculation method [54] was chosen to measure the quantities of unclear materials for calculation simplification, such as foundations and steel bars in the concrete structure. A 5% and 10% surplus for different materials were added in order to account for on-site processes, broken elements, and purchased quantities.

3.2. Implementation of Sensitivity Analysis in Building Models

Sensitivity analysis (SA) is the observational study of identifying the most influential input parameters for the output behaviour in the model. Besides increasing the understanding of the relationships between design parameters and objectives of key performance indicators (KPIs), SA can be a valuable tool for supporting a design–optimization process, by narrowing the search ranges and limiting the number of factors to only the most significant ones. Therefore, it has been broadly used in the domain of building performance analysis and explores the highly sensitive parameters influencing energy performance for design support in various building types or climate zones [55,56]. Two major methods for SA are termed 'local' and 'global' as summarized in [57]. The global sensitivity analysis (GSA), as a more reliable method, includes regression (e.g., SRC), screening methods (e.g., Morris), and variance based (e.g., FAST, Sobol indices) and meta-model approaches. When restricted to the quantification of environmental impact, it behaves linearly, with standardized regression coefficients (SRC), etc., that can be chosen [57]. Another technique to avoid extensive combinations of extreme values is stochastic modelling with the aid of a Monte Carlo simulation or Latin hypercube sampling (LHS). In the simulations, a predefined, limited number of combinations of random parameters are used to calculate the outcomes [58].

To simplify the complexity of evaluation, in this study a SA of multiple design variables of building envelope in software to operational GHG emissions was performed to evaluate the highly sensitive parameters. The results are used for preliminary judgment, to predict optimal design strategies, and to conduct further analysis. Design strategies in terms of space and material could not be defined as design variables in aided tools; therefore, the analysis was conducted separately and manually. The SA method implemented in this study is SRC, with the sampling method of LHS. SRC value, as a sensitivity index, is used to describe the relative influence of variability parameters on the output results of a model. A high value of SRC indicates that the parameter has a strong influence on the results. The output objective of 'operational CO_{2eq} emissions' is defined and calculated by referring to annual total energy related emissions.

3.3. BIM-LCA Approach

For an efficient and systematic connection of building models, energy simulation, and life cycle GHG calculation, this study integrated all work based on a 'building information model (BIM)' with support tools. The case model, built in Autodesk Revit, contains the building model's geometric, space, and some thermal information; therefore, it is referred to as the 'building information model'. For building energy simulation and further design optimization, the 'building information model' in Autodesk Revit was then converted to the 'building energy model' directly, using simulation software DesignBuilder with EnergyPlus engine. Some complicated and special components should be simplified for making it efficient to run a simulation, but close to the real situations. For instance, beams and columns of structures are deleted and some curves of window corners are changed to straight lines. The energy used for operation of the building was calculated through specific input data for energy simulation. An Excel based LCA tool was used in this study, which was developed by the Research Centre for Zero Emission Buildings and verified in life cycle GHG emission calculations for numerous case studies and pilot buildings [10,11,59]. Proper categories and boundaries are set in order to acquire relevant outputs. Quantities of different materials and components have been exported from the Revit/BIM model, and then delivered to the LCA tool for detailed GHG calculations. The materials and components contained in the BIM output correspond to the categories in the LCA tool. In DesignBuilder, GSA of operational energy emissions was also conducted to filter out the highly sensitive parameters, which are the key design strategies of the building envelope to be considered.

4. Case-TJS

4.1. Case Description

Almost all medium-sized HSRS buildings in the cold zone follow the characteristics mentioned in Section 2.1.2. Additionally, they have a similar structure and envelope pattern and thus the variables are limited. The materials used for the building structure were reinforced concrete, and the envelope is mostly hollow concrete block with a curtain wall system. Case-TJS is used as a representative sample in this study. The results of the analysis are applicable to other medium-sized HSRS buildings in China.

Case-TJS is a 2-story building comprising a ground floor (0F) and a first floor (1F, platform), covering a GFA of 8669 m^2. It is located in the Xiqing District, southwest of Tianjin, China. The Xiqing district is at 39°N latitude with a mean daily outside relative humidity of 62–73%. The mean daily outside air temperature is 29.3 °C in summer and −9.4 °C in winter. The case building is composed of two parts, the west wing building (rail side) and east wing building below the railway lines or platform (rail below). Typically, the west wing building is representative in space, structure, envelope, and materials. The internal height is 30 m, the same as the large waiting room. The width and depth are 145 m and 20 m, respectively. Case-TJS is oriented east west, as are other stations in the Beijing–Shanghai railway network. The maximum assembling passenger number is about 1000 people.

The structural components refer to the column and beams; the foundation and the floor slab are composed of reinforced concrete, others being steel truss structures in the west wing roof. The external walls contain a small concrete hollow block, 100 mm Rockwool (west wing) or 60 mm polystyrene (east wing) insulation with an ETERPAN fibre cement board. A double low-E glazing glass curtain wall was used as a waiting hall envelope. The outer roof of the west wing consists of 100 mm of aluminium foil glass wool insulation with colour steel lamellas for the sunscreen. The east wing roof is mainly covered with a reinforced concrete slab and 80 mm extruded polystyrene (XPS) insulation. No skylights and solar thermal collectors are currently installed on the roof. The floor is decking with 30 mm of XPS insulation, fine stone concrete, and granite.

4.2. Parameter Setting for Simulation

Figure 5a shows the building information model built in Autodesk Revit. Figure 5b shows the building energy model in the simulation software DesignBuilder. The meteorological data used for the simulation correspond to the local climate, according to the Chinese Standard Weather Data. Heated rooms, such as the waiting hall and office, are the primary areas with floor heating and XPS insulation. The indoor design parameters of occupancy, the HVAC system, lighting, and domestic hot water (DHW) are in accordance with the real condition and code for design of a railway passenger station [39], and are shown in Table 3. The waiting hall/room is a typical feature in station buildings; the parameter in these areas is particularly important. The occupancy density of the waiting hall is determined as 0.12 persons/m^2 considering the high mobility and regularity of passengers. To achieve satisfactory levels of thermal comfort in summer, most railway stations in cold regions have a central air conditioning system. Electric power is needed for energy supply solutions. The building has a balanced mechanical ventilation system equipped with a heat recovery unit. The variable air volume ventilation operates a mixed flow in the waiting and ticket room: the fresh air is supplied into the building from the top and is extracted down into devices below. Standard hygienic flow rates are imposed, with minimum fresh air 2.778 l/s per person. The building is equipped with electric water to the water/ground source heat pump (GSHP) to cover the thermal needs (i.e. space heating and cooling). A floor heating system is used in the waiting room. The value of the whole system seasonal coefficient of performance (SCOP) in heating and the energy efficiency ratio (SEER) in cooling for specific pump power is set to 4.0 and 4.8 respectively. The heat pump produces hot water with a temperature of 45 °C in winter, and cold water combined with air-to-water chillers in summer (supply temperature of 12 °C). The energy consumption of fans, pumps and other auxiliary equipment is calculated. The consumption rate of 0.4 l/m^2·day is supplied for DHW with a draw-off temperature set at 65 °C. These reflect the practices in modern medium-sized HSRS buildings. The operation schedule is from 05:00 to 23:00, in keeping with the existing train time of day scheduling in Case-TJS. The indoor illuminance of the waiting room is controlled to be 200 lux and the power density is 8 W/m^2.

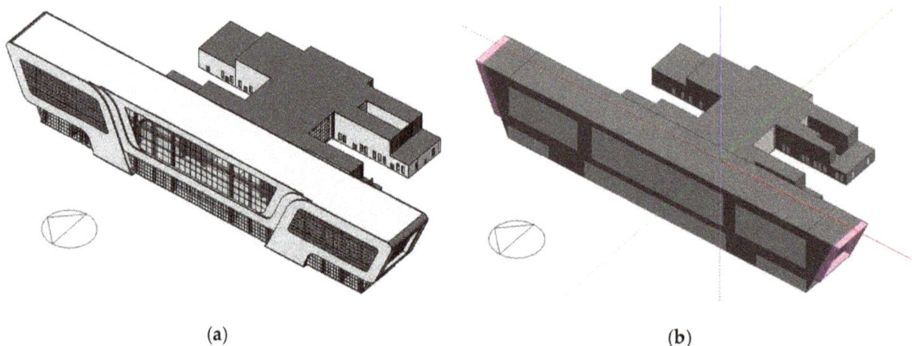

(a)   (b)

Figure 5. (**a**) Bird's eye view of building information model (BIM) in Revit; (**b**) Building energy model in DesignBuilder.

Table 3. Design parameters for the occupancy, HVAC system, lighting, and DHW.

Design Parameters		Values
Occupancy Density (Persons/m^2)		**0.12 (Waiting Room)**
	Heating design temperature	18 °C (waiting room)
Heating	SCOP	4.0 (GSHP, electricity)
	Operating schedule	15th November–15th March; 5:00–23:00 on
	Cooling design temperature	28 °C (waiting room)
Cooling	SEER	4.8 (GSHP, electricity)
	Operating schedule	15th May–15th September; 5:00–23:00 on
HVAC system	Mechanical ventilation–air exchange	2–3/h
	Minimum fresh air (l/s·person)	2.778
	Illuminance (lux)	200 (waiting room)
Lighting efficiency	Power density (w/m^2)	8 (waiting room)/6 (others)
	Operating schedule	5:00–23:00 every day, whole year
DHW	Consumption rate (l/m^2·day)	0.4

5. Results and Discussion

5.1. LCCO$_{2eq}$ by Life Cycle Module

According to energy simulation results shown in Figure 6, the total annual electricity consumption for the building is 706,523 kWh/year, including heating, cooling, lighting, fans and pumps, and DHW. The operation energy use per unit area of the case building is 81.5 kWh/m^2·a, corresponding to operational energy emissions of 94 kgCO$_{2eq}$/m^2·a, as electric CO$_{2eq}$ factor is 1.15 kgCO$_{2eq}$/kWh. Heating and cooling are the main energy consumers in Case-TJS, followed by lighting, which is consistent with the previous study about large-sized HSRS buildings [22].

Figure 7 shows the contribution of different life cycle modules in GHG emissions, assuming a 50-year life span. The total LCCO$_{2eq}$ is calculated to be 115 kgCO$_{2eq}$/m^2·a. The majority of emissions apparently occur during the operational energy use phase (O, or B6), constituting a percentage of 81%. Thus, operational energy emissions are much more dominant than embodied GHG emissions for the case study. This can mainly be due of the high emission factor used for the grid. As a previous analysis shows, low utilization of vertical height is perhaps also a reason for not only high energy consumption and associated emissions in domestic HSRS buildings, but also low occupant density of the waiting room. Construction material emissions from the production and replacement phases (A1–A3 and B4) contribute 14.1% and 4.7% respectively, with a high level of 22 kgCO$_{2eq}$/m^2·a in total, compared to other building typologies. The embodied GHG emissions of life cycle module A1–A3 is 3.0 times that of module B4. The calculation results also highlight the significance of embodied GHG emissions, accounting for approximately 19% of GHG emissions generated from the life cycle of HSRS buildings.

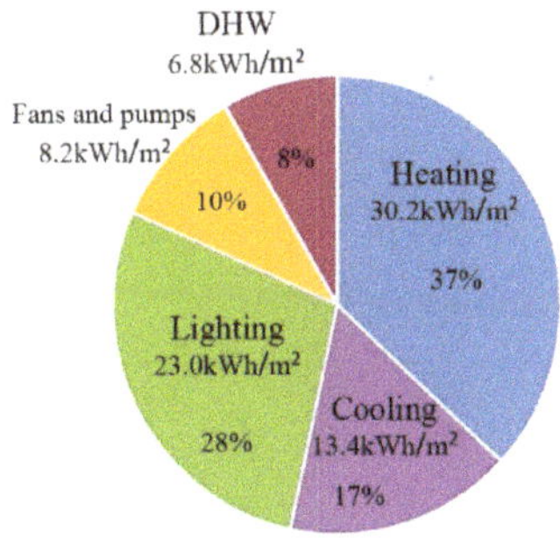

Figure 6. Distribution of annual operational energy use (kWh/m^2·a).

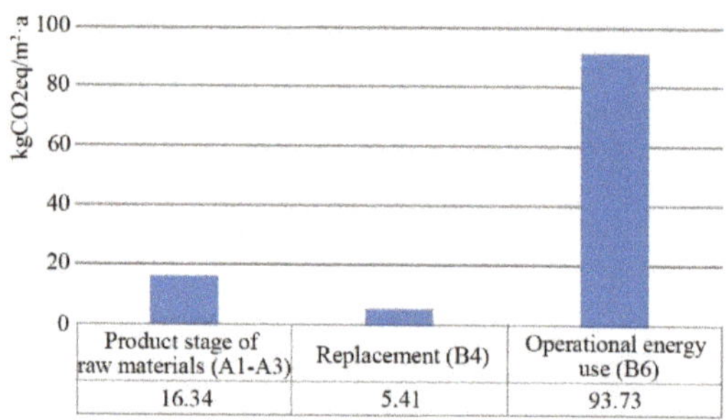

Figure 7. GHG emissions per life cycle module (kgCO$_{2eq}$/m^2·a).

5.2. Embodied GHG Emissions of Building Materials

5.2.1. Embodied GHG Emissions by Building Elements

This study detailed the embodied GHG emissions of each section in life cycle modules A1–A3 and B4. The embodied GHG emissions of different building elements and the percentage they constitute are summarized in Table 4. The results show that outer walls are the largest contributor to high embodied GHG emissions across the building elements, contributing approximately 48% to total embodied GHG emissions; the load bearing structure contributes approximately 26%; the floor structure and outer roof contribute approximately 10% equally; while groundwork, foundations, and inner walls contribute the least. A breakdown of embodied GHG emissions by each building element is shown in Figure 8. Outer walls and load-bearing structures are main drivers for high embodied GHG emissions because they constitute the main construction elements of the building and use materials with high embodied GHG emissions such as reinforced concrete and steel.

Table 4. GHG emissions per building element and the proportion that they constitute.

	Life Cycle Module A1–A3		Life Cycle Module B4		Life Cycle Module A1–A3, B4	
	CO$_{2eq}$ Emissions (kgCO$_{2eq}$/m^2·a)	Percentage (%)	CO$_{2eq}$ Emissions (kgCO$_{2eq}$/m^2·a)	Percentage (%)	CO$_{2eq}$ Emissions (kgCO$_{2eq}$/m^2·a)	Percentage (%)
Groundwork, foundations	0.58	3.56	0	0	0.58	2.67
Load bearing structure	5.65	34.61	0	0	5.65	25.98
Outer walls	6.19	37.92	4.17	77.13	10.37	47.68
Inner walls	0.60	3.69	0	0	0.60	2.76
Floor structure	1.68	10.29	0.47	8.77	2.16	9.93
Outer roof	1.62	9.92	0.76	14.10	2.38	10.94
Total	16.34	100	5.41	100	21.75	100

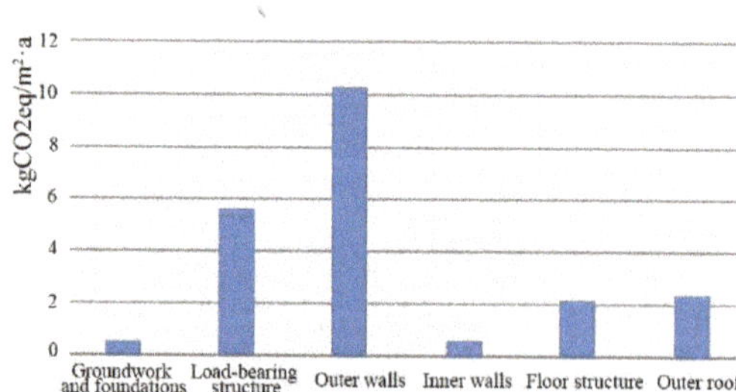

Figure 8. Comparison of embodied GHG emissions by building elements.

5.2.2. Embodied GHG Emissions by Construction Materials

Figure 9 shows the embodied GHG emissions by major construction materials. The blue histogram represents emission results from material production phases (A1–A3) of each material and the proportion they constitute out of all the materials, while the orange histogram represents emission results from the material replacement phase (B4). By analyzing the blue histogram, the highest GHG emissions are generated from three types of construction materials—concrete, steel, and insulation materials—contributing nearly 71% totally to GHG emissions of material production. Of these emissions, more than half (ca. 52% total) are generated from concrete and steel materials, which are the design drivers. For an HSRS building, a large waiting room space means a large span and greater height. The former requires higher structural strength and will lead to high use of steel material from steel roof and reinforced concrete structure, such as load bearing (architectural beam and column). The concrete consumption will increase to support the greater height as well.

However, when comparing the GHG emissions that belong to material replacement (see orange histogram in Figure 9), a big difference in trend can be found, as follows: the high GHG emissions of material replacement are mainly from three types of construction materials—insulation material, windows, and plaster—among which insulation material is the design driver. This distribution is caused by the lower service life of these materials: 25 years. Thus, during the buildings' 50-year life span, insulation materials, windows, and plaster are required to be replaced once. If comparing the total embodied GHG emissions (A1–A3, B4), the insulation emission would be much greater than that of other materials.

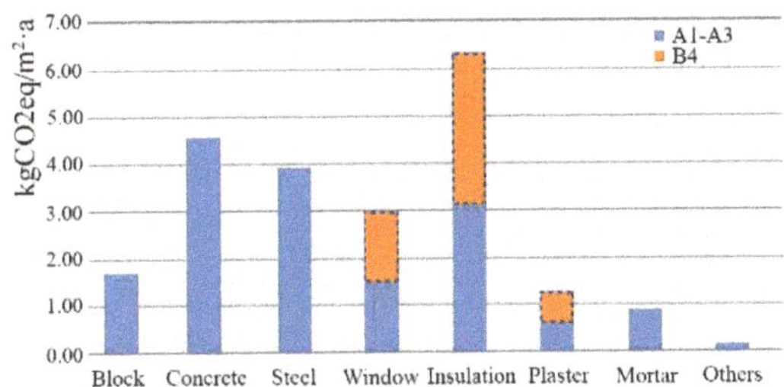

Figure 9. Level of material inventory detail and comparison of embodied GHG emissions.

5.3. Reduction Strategies of Life Cycle GHG Emissions

The case originally aimed at evaluating $LCCO_{2eq}$ to extract design drivers and influences. It is believed that high emissions are affected by three aspects of buildings: space, envelope, and materials. Due to stringent energy requirements of design standard and energy efficiency improvement [38], the reduction possibilities in operational energy emissions by means of an energy system are obvious. The building energy consumption of the case study, indeed, is lower compared to that of other conventional railway station buildings [20], mainly due to the usage of the high performance system heat pump. However, in this article, the main attention has been paid to the building performance at the design level (or, passive strategies) rather than service systems such as HVAC. A building envelope—especially outer wall, roof, and glass curtain wall—has an impact on the energy performance of buildings and the main building elements for GHG calculation. Therefore, for an HSRS building as specific typology, reduction possibilities and potentials of $LCCO_{2eq}$ should be identified on building space, building envelope, and material, as the scope of design strategies.

5.3.1. Reduction Strategies of Operational GHG Emissions

Optimizing Building Space

The space of Case-TJS was optimized in two ways: changing the height and area of the waiting room. The height of the west wing building was reduced from 30 m, 25 m, to 20 m, itemized as strategy H_1, H_2, and H_3, respectively. The operational GHG emissions decrease by 15.7% and total $LCCO_{2eq}$ decreases by 15.0% from strategy H_1 to H_3. Embodied GHG emissions decrease by up to 12.3% and see a significant reduction. This is because, with the decline of the building height, the enclosed volume to use energy for heating and cooling is reduced. Meanwhile, fewer materials of outer walls and the load bearing structure are consumed. The area of the waiting room was reduced from 2346 m^2, 1313 m^2, to 657 m^2 through functional adjustment, itemized as strategy A_1, A_2, and A_3, respectively. It also resulted in more enclosed space being opened without heating and cooling for passengers going through the building to the platform in half an hour. Due to the fact that the area of the waiting room is determined by the maximum number of passengers (H), this change assumed that 0%H, 36%H, and 72%H can check in directly, while others remain in the enclosed waiting room. Although embodied GHG emissions remain unchanged, the operational GHG emissions decrease greatly, by up to 24.8%, leading to a decrease in the total GHG emissions by 20.1%.

Optimizing Building Envelope

The design variables are quantified using a computational tool, combining options into 256 samples (4 × 4 × 4 × 4). From the SA results shown in Table 5, it can be determined that the glazing type has the most correlative influence on operational GHG emissions, with the largest SRC value, followed by external walls and outer roof constructions. By linear regression analysis results together with SA, deterministic coefficient adjusted R^2 of regression equation reaches 0.97, which indicates that there is a strong linear relationship between design variables and operational GHG emissions. After correlation and regression analysis, glazing type was selected as an optimal design variable of envelope to make changes for emission reduction. The double low-E glazing type of curtain walls (8 + 12Air + 8) and windows (6 + 9Air + 6) was changed in the case building to double low-E coated glazing (6 + 12Air + 6) and double low-E insulated glazing (6 + 12Argon + 6), with different specific parameters (U-value, Solar Heat Gain Coefficient—SHGC), itemized as strategy U_{G1}, U_{G2}, and U_{G3}. As a result, the impacts decrease by 7.7% during the operational phase from strategy U_{G1} to U_{G3} because of the change with high performance and low emissivity materials. Embodied GHG emissions decrease by 6.8%. In addition, an optimal solution is taken from the samples simultaneously (see Table 5), which could be used in the comprehensive comparison of different optimization scenarios with multiple strategies.

Table 5. Design variables matrix for SA, SRC values, and optimal solution.

	Design Variables	Range/Values	Number of Design Options	SRC Value	Case Building	Optimal Solution
1	External wall construction (insulation thickness)	0.5 cm (U = 0.56 W/m^2·K)	4 options	0.42		
		1.0 cm (U = 0.42 W/m^2·K)			•*	
		2.0 cm (U = 0.27 W/m^2·K)				
		3.0 cm (U = 0.20 W/m^2·K)				•
2	Outer roof construction (insulation thickness)	0.5 cm (U = 0.64 W/m^2·K)	4 options	0.39		
		1.0 cm (U = 0.33 W/m^2·K)			•	
		2.0 cm (U = 0.17 W/m^2·K)				•
		3.0 cm (U = 0.12 W/m^2·K)				
3	Glazing type (U-value)	Double low-E insulated glazing (U = 1.4 W/m^2·K, SHGC = 0.30)	4 options	0.87		•
		Double low-E coated glazing (U = 1.7 W/m^2·K, SHGC = 0.50)				
		Double low-E glazing (U = 2.2 W/m^2·K, SHGC = 0.63)			•	
		Double clear float glazing (U = 2.8 W/m^2·K, SHGC = 0.70)				
4	Local shading (projection louver)	0	4 options	0.19	•	
		0.5 m				
		1.0 m				
		1.5 m				•

* dot denotes the corresponding value in "Range/values" column that was used in case building or selected in optimal solution.

5.3.2. Reduction Strategies of Embodied Emissions

Construction materials could be optimized with regard to embodied emissions, by alternative building design and material solutions. In life cycle GHG calculations, it is assumed that material alternatives only change emission data or the service life of materials, while technical properties such as load bearing capacity and thermal performance remain the same as before.

Substitution of Construction Materials

The above analysis reveals that concrete and steel are the main contributors for embodied GHG emissions from materials. No recycled or reused materials are used in reality, itemized as strategy S_1 for comparison. In reduction strategies, low embodied carbon materials are chosen. As a reference, railway stations can be evaluated at a good level in material utilization if the percentage weight of recycled and reused materials is more than 20% of that of total construction materials [60]. To reach the 20% target, recycled aggregate concrete and recycled steel were substituted for 30% of reinforced concrete and structural steel in the case building, itemized as strategy S_3. It represents the weight of recycled aggregate concrete and recycled steel, constituting 20% of the total amount of concrete and steel. In addition, 15% of reinforced concrete and structural steel were substituted by recycled aggregate concrete and recycled steel, itemized as strategy S_2. Incorporating the recycled concrete and steel in calculations can help reduce embodied GHG emission impacts arising from materials, although there is little influence on total GHG emissions (a 0.7% reduction).

Service Life Extension of Insulation Material

Maximizing the use of materials can reduce the number of times replacements are required during building operation; therefore, it has a great effect on replacement. The service life of insulation increased from 25 years to 35 years and 50 years compared to the case building, itemized as strategy E_1, E_2, and E_3. The result reveals that embodied GHG emissions greatly decrease up to 15%, while total GHG emissions reduce by 2.8% after extending the service life of materials from 25 years to 50 years. Single strategies optimized in the case building and GHG emissions are presented in Table 6 and the histogram given in Figure 10.

Table 6. Single strategies optimized in the case building and GHG emissions ($kgCO_{2eq}/m^2{\cdot}a$).

Single Strategies	Embodied GHG Emissions			Operational GHG Emissions	Total Life Cycle GHG Emissions
	Module A1–A3	Module B4	Total Modules		
Case building ($H_1/A_1/U_{G1}/S_1/E_1$)	16.34	5.41	21.75	93.73	115.48
$H_2 = 25$ m	15.58	5.06	20.64	83.54	104.18
$H_3 = 20$ m	14.63	4.45	19.08	78.98	98.06
Reduction (%) *	−10.5%	−17.7%	−12.3%	−15.7%	−15.0%
$A_2 = 1313$ m^2	16.34	5.41	21.75	76.39	98.14
$A_3 = 657$ m^2	16.34	5.41	21.75	70.46	92.21
Reduction (%) *	-	-	-	−24.8%	−20.1%
$U_{G2} = 1.7$ W/m$^2{\cdot}$K	15.79	4.86	20.65	90.15	110.80
$U_{G3} = 1.4$ W/m$^2{\cdot}$K	15.60	4.67	20.27	86.51	106.78
Reduction (%) *	−4.5%	−13.6%	−6.8%	−7.7%	−7.5%
$S_2 = 15\%$ recycled concrete + steel	15.93	5.41	21.34	93.73	115.07
$S_3 = 30\%$ recycled concrete + steel	15.52	5.41	20.93	93.73	114.66
Reduction (%) *	−5.0%	-	−3.7%	-	−0.7%
$E_2 = 35$, service life	16.34	5.41	21.75	93.73	115.48
$E_3 = 50$, service life	16.34	2.15	18.49	93.73	112.22
Reduction (%) *	-	−60.2%	−15.0%	-	−2.8%

* denotes the GHG emissions reduction (%) of building optimized by single strategies (H_3, A_3, U_{G3}, S_3, E_3) compared to the reference (H_1, A_1, U_{G1}, S_1, E_1) in each column.

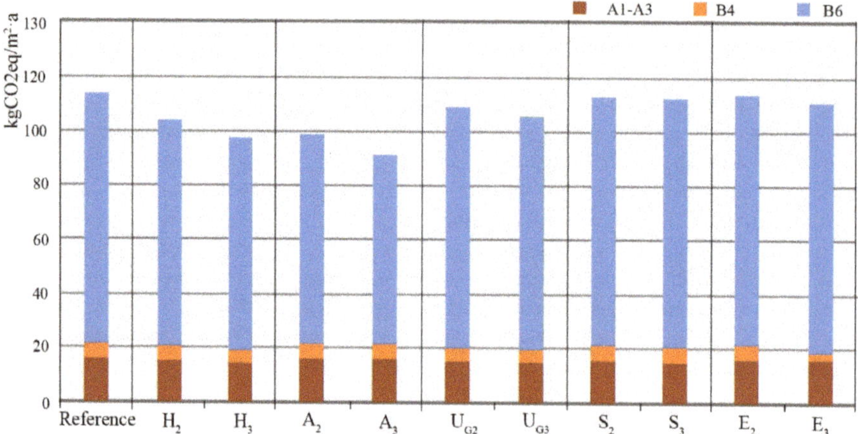

Figure 10. GHG emissions of case building with different single strategies.

5.3.3. Reduction Potential Analysis of Life Cycle GHG Emissions

In order to evaluate the reduction potential comprehensively, three optimization scenarios with multiple strategies were proposed under new simulation and calculation: space, envelope, and material optimization. Space optimization corresponds with the optimal single strategy such as H_3 and A_3 that were integrated in the case. Envelope optimization corresponds to an optimal solution in Table 5 that was integrated in the case. Optimal single strategy S_3 and E_3 were integrated as material optimization. Figure 11 shows the annual operation energy use in all three scenarios. In scenario 1, the operational energy use of the case building was reduced to 50–60 kWh/m²·a, which is close to the median value in offices or schools shown in Figure 1. A comparison was made between the total GHG emission results of optimization scenarios, as summarized in Table 7. Taking the origin case as a reference, it can be seen that space optimization has the greatest reduction potential of total GHG emissions, with a reduction percentage of 28.2%. There are fewer changes relatively in total GHG emissions under material optimization, but it contributes to a 3.5% reduction. In this study, if these three optimizations are all integrated into the case with model adjustments and updated parameters, the total GHG emissions will decrease up to 33.8% after recalculation.

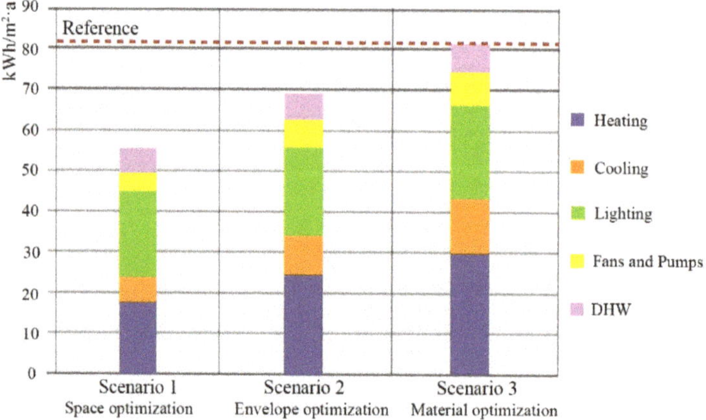

Figure 11. Annual operation energy use in all three scenarios (kWh/m²·a).

Table 7. Summary of the life cycle GHG emission results of different optimization scenarios ($kgCO_{2eq}/m^2{\cdot}a$).

	Space Optimization (H_3, A_3 in Table 1)	Envelope Optimization (Optimal Solution in Table 1)	Material Optimization (S_3, E_3 in Table 1)
A1–A3	14.63	15.60	15.52
B4	4.45	4.67	2.15
B6	63.80	80.01	93.73
Embodied GHG emissions	19.08	20.27	17.67
Total GHG emissions	82.88	100.28	111.4
Reduction (%) *	−28.2%	−13.1%	−3.5%

* denotes the reduction value (%) of total GHG emissions in optimization scenarios compared to the reference value ($115\ kgCO_{2eq}/m^2{\cdot}a$).

There are key reductions in operational GHG emissions when space form varies, indicating that optimizing building space has great potential for reducing energy associated GHG emissions, especially by reducing the area of enclosed waiting rooms. Improving the waiting behaviour of passengers is significant for emission reduction on the assumption that more passengers can check in directly. The result is consistent with the previous study [61], which indicates the key link between the operation phase and emissions reduction for similar large space public buildings such as museum buildings in China. Compared with a Swedish HSRS building [27], the result is reversed: construction and maintenance are more dominant in GHG emissions, while operation gives just a small contribution as a result of used green electric power. Therefore, it is meaningful to study HSRS buildings in different regions.

It should be noticed that the linear relationship between operational GHG emissions and area is not evident, as investigated in [16] the regarding correlation between energy consumption and station area. In energy simulations of this study, energy use from the waiting room accounts approximately for 60% of total energy consumption from all operating rooms. Although the energy use of the waiting room could greatly decrease, the energy consumed in other rooms including the booking hall, offices and commercials remains unchanged. Lighting and DHW constitute 36% in annual operational energy use, but they are not sensible when waiting room area changes are a necessary service for passengers in semi-open areas. This is why operational GHG emissions reduce only by 24% when the area of the waiting room reduces drastically from 2346 m^2 to 657 m^2 by 72%.

Another focus should be on using low emissivity and high performance materials/components, e.g., glazing type. In terms of embodied GHG emission reduction, extending the service life of insulation material contributes the most, followed by optimizing space. Using recycled concrete and steel has a relatively minimal impact. In practice, recycled materials are usually insufficient and the amount for substitution is limited. In order to minimize embodied GHG emissions, efforts should be made to choose robust insulation materials (i.e. longer service life) and to reduce the amount of materials. Recycled concrete and steel with low emissions could also be used to make contributions more or less.

In three scenarios, space optimization can greatly reduce operational as well as embodied GHG emissions, indicating the significance of space design in emission reduction during the architectural design process. The reduction potential of total GHG emissions from envelope optimization comes from the performance improvement in terms of thermal properties and low emissivity. However, this improvement is limited because newly built HSRS buildings have up to or over the required standard values on envelope [38], whereas stricter performance costs more. Generally, compared to the innovative incorporation of state-of-the-art materials and technology to drive down emissions relating to the operational phase, the process of space optimization in buildings is deemed time saving and is economically applicable.

As life cycle GHG emissions calculations are developed, the robustness of these results against uncertainties should be noticed. The choice of electricity grid factor has a large influence on operational GHG emissions. If assuming factor value of the current grid of 0.361 $kgCO_{2eq}/kWh$ in the EU [62], the operational GHG emissions could be close to the embodied GHG emissions. The future studies should test different scenarios with the decarbonization of the electricity grid over building lifespan. The life cycle analysis presented in this study also involves a number of simplifications and generalisations. The emission factors of construction materials use generic data from standard and research databases, although these factors may vary in a specific practice. The construction materials have been limited to civil engineering excluding building services. Results are therefore discussed for these specific contexts. Although simplified, this framework of analysis is nevertheless representative for most of the medium-sized HSRS buildings in the cold zone.

The proposed BIM-LCA method connects LCA with BIM and comprehensively applies different software for study systematically, however, there are some limitations. Although the simulation software DesignBuilder has improved abilities to analyze the Revit models, some incompatibility problems occur, especially when the model is established in a more complex way. Non-linear geometric primitives should be simplified. For the Case-TJS, most of the origin components in the Revit model have been imported into Designbuilder for smoother energy analysis, which plays a positive role in the BIM-LCA method.

6. Conclusions

Previous research usually does not pay sufficient attention to HSRS buildings from the direct perspective of life cycle GHG emissions and how to reduce GHG emissions. Therefore, this article set out to evaluate and present the life cycle GHG emissions ($LCCO_{2eq}$) of an HSRS building in a cold zone, China to extract the design drivers in terms of operational and embodied GHG emissions. Design drivers and reduction strategies were examined in order to assess their potential as strategies for reduced GHG emissions from a functional and architectural point of view. This paper proposed a detailed methodology for the development and assessment of emission reduction strategies through LCA, combined with BIM technology. The Case-TJS, a medium-sized HSRS building, was selected as a representative example for a case study. This study comes to the following conclusions:

(1) Emission reduction measures in an HSRS building should focus more on space design in the early stage of architectural design. Although the GHG emission mitigation strategy related to the substitution of materials presents the lowest potential for total GHG emissions reduction, significant embodied emission reduction can be achieved by choosing insulation materials with longer service life.

(2) The $LCCO_{2eq}$ of an HSRS building for design analysis were assessed using 3 system boundaries: life cycle modules A1–A3 and B4 for the production and replacement of building materials, and life cycle module B6 for operational energy use. The BIM-LCA approach shows how modelling tools help analysis identify in GHG reduction strategies in complex buildings such as HSRS.

(3) The main objective of the GHG reduction strategies in HSRS buildings is to minimize the total GHG emissions related to operational energy and, ultimately, the embodied emissions from materials.

(4) The drivers for the highest embodied GHG emissions were from concrete, steel, and insulation materials used in the main load bearing structures and outer wall components.

In summary, this paper contributes to research by identifying strategies to reducing GHG emissions with a focus on the HSRS building typology which is a particular type of large space public building in China. It is anticipated that the research results that might significantly reduce GHG emissions in HSRS buildings would have a positive impact on the mitigation of current climate change effects. The results for GHG emissions and reduction strategies also provide guidance to help inform design and construction decisions of similar projects and large space public buildings.

Author Contributions: Conceptualization, N.W.; methodology, N.W.; validation, N.W. and D.S.; formal analysis, N.W.; investigation, N.W.; resources, C.L., A.H.W. and D.S.; writing—original draft preparation, N.W.; writing—review and editing, D.S., C.L. and A.G.; visualization, N.W.; supervision, A.H.W., C.L. and A.G.; project administration, C.L. and A.G.; and funding acquisition, C.L. All authors have read and agreed to the published version of the manuscript.

Funding: This research was funded by the National Key Research and Development Program of China (Grant No. 2016YFC0700200), National Natural Science Foundation of China (Grant No. 51808383), and the Programme of Introducing Talents of Discipline to Universities (project No. B13011).

Acknowledgments: The authors would like to thank the China Scholarship Council (CSC) for providing financial support to carry out the cooperation study. The authors gratefully acknowledge the support from the Research Council of Norway and several partners through the Research Centre on Zero Emission Neighbourhoods in Smart Cities (FME ZEN), hosted by the Norwegian University of Science and Technology (NTNU).

Conflicts of Interest: The authors declare no conflict of interest.

References

1. International Energy Agency and the United Nations Environment Programme. *Global Status Report 2018: Towards a Zero-Emission, Efficient and Resilient Buildings and Construction Sector*; International Energy Agency and the United Nations Environment Programme: Katowice, Poland, 2018; Available online: https://www.unenvironment.org/resources/report/global-status-report-2018 (accessed on 15 September 2019).

2. Department of Climate Change, National Development and Reform Commission of China. *Enhanced Actions on Climate Change*; China's Intended Nationally Determined Contributions (INDC): Beijing, China, 2015.

3. State Council of China. 13th Five-Year Plan of Energy Saving and Emission Reduction. Available online: http://www.gov.cn/zhengce/content/2017-01/05/content_5156789.htm (accessed on 20 September 2019).

4. Building Energy Efficiency Centre, Tsinghua University. *Annual Report on Development of China Building Energy Efficiency*; China Architecture and Building Press: Beijing, China, 2019.

5. Wang, H.; Zhou, P.; Guo, C.; Tang, X.; Xue, Y.; Huang, C. On the calculation of heat migration in thermally stratified environment of large space building with sidewall nozzle air-supply. *Build. Environ.* **2019**, *147*, 221–230. [CrossRef]

6. Building Energy Efficiency Centre, Tsinghua University. *Annual Report on Development of China Building Energy Efficiency*; China Architecture and Building Press: Beijing, China, 2014.

7. National Development and Reform Commission of China. *The Mid-Long Term Planning for China's Railway Network (2016–2025)*; National Development and Reform Commission of China: Beijing, China, 2016.

8. Luo, Z.; Yang, L.; Liu, J. Embodied carbon emissions of office building: A case study of China's 78 office buildings. *Build. Environ.* **2016**, *95*, 365–371. [CrossRef]

9. Zhang, X.; Wang, F. Life-cycle assessment and control measures for carbon emissions of typical buildings in China. *Build. Environ.* **2015**, *86*, 89–97. [CrossRef]

10. *The Research Centre on Zero Emission Buildings (ZEB), Zero Emission Buildings, NTNU and SINTEF*; Fagbokforlaget: Bergen, Norway, 2017.

11. Wiik, M.K.; Fufa, S.M.; Kristjansdottir, T.; Andresen, I. Lessons learnt from embodied GHG emission calculations in zero emission buildings (ZEBs) from the Norwegian ZEB research centre. *Energy Build.* **2018**, *165*, 25–34. [CrossRef]

12. Rodríguez Serrano, A.Á.; Porras Álvarez, S. Life Cycle Assessment in Building: A Case Study on the Energy and Emissions Impact Related to the Choice of Housing Typologies and Construction Process in Spain. *Sustainability* **2016**, *8*, 287. [CrossRef]

13. Amirkhani, S.; Bahadori-Jahromi, A.; Mylona, A.; Godfrey, P.; Cook, D. Impact of Low-E Window Films on Energy Consumption and CO_2 Emissions of an Existing UK Hotel Building. *Sustainability* **2019**, *11*, 4265. [CrossRef]

14. French, J. SusStations Project. In Proceedings of the 7th International Conference on. Improving, Energy Efficiency in Commercial Buildings (IEECB), Frankfurt, Germany, 18–19 April 2012.

15. Indian Green Building Council (IGBC), IGBC Green Railway Stations. Available online: https://igbc.in/igbc/redirectHtml.htm?redVal=showGreenRailwaynosign (accessed on 5 October 2019).

16. Song, L.; Wang, Y.; Li, X. Energy performance and environmental quality of typical railway passenger stations in northern China. *Indoor Built Environ.* **2018**, *27*, 296–307. [CrossRef]

17. Yang, X.E. Building energy efficiency design for waiting halls of railway stations in the hot summer and cold-winter region. *J. Eng. Sci. Technol. Rev.* **2017**, *10*, 1–7. [CrossRef]

18. Yang, L.; Xia, J. Case study of space cooling and heating energy demand of a high-speed railway station in China. *Procedia Eng.* **2015**, *121*, 1887–1893. [CrossRef]

19. Gang, L.; Yi, X.; Rui, D.; Chen, L. The research of triadic relation among building spaces, lighting comfort level and lighting energy consumption in high-speed railway station in China. *Procedia Eng.* **2015**, *121*, 854–865. [CrossRef]

20. Asian Development Bank. *Improving Energy Efficiency and Reducing Emissions through Intelligent Railway Station Buildings*; Asian Development Bank: Mandaluyong, Philippines, 2015.

21. Wang, N.; Wang, J.L.; Liu, C.H.; Liu, L. Energy-Saving Potential of Large Space Public Buildings Based on BIM: A Case Study of the Building in High-Speed Railway Station. In *eWork and eBusiness in Architecture, Engineering and Construction*; Karlshoj, J., Scherer, R., Eds.; CRC Press: Copenhagen, Denmark, 2019. [CrossRef]

22. Wang, J.; Wang, N.; Liu, L.; Liu, C. Energy-efficient analysis of high-speed railway station design in the cold region: A case study on Tianjin West Railway Station. *Eco-City Green Build* **2018**, *9*, 44–51. (In Chinese)

23. Li, Q.; Yoshino, H.; Mochida, A.; Meng, Q.; Lei, B.; Zhao, L.; Lun, Y. CFD study of the thermal environment in an air-conditioned train station building. *Build. Environ.* **2009**, *44*, 1452–1465. [CrossRef]

24. Deb, C.; Ramachandraiah, A. Evaluation of thermal comfort in a rail terminal location in India. *Build. Environ.* **2010**, *45*, 2571–2580. [CrossRef]

25. Chow, W.K.; Fung, W.Y.; Wong, L.T. Preliminary studies on a new method for assessing ventilation in large spaces. *Build. Environ.* **2002**, *37*, 145–152. [CrossRef]

26. Chong, U.; Swanson, J.J.; Boies, A.M. Air quality evaluation of London Paddington train station. *Environ. Res. Lett.* **2015**, *10*, 1–11. [CrossRef]

27. Stripple, H.; Uppenberg, S. *Life-Cycle Assessment of Railways and Rail Transport—Appplication in Environmental Product Declarations (EPDs) for the Bothnia Line*; Swedish Environmental Research Institute: Göteborg, Sweden, 2010; Available online: https://www.ivl.se/download/18.343dc99d14e8bb0f58b75d4/1445517456715/B1943.pdf (accessed on 3 February 2020).

28. Zhang, R.; Liu, P.; Zhou, C.C.; Amorelli, A.; Li, Z. Configuration of inter-city high-speed passenger transport infrastructure with minimal construction and operational energy consumption: A superstructure based modelling and optimization framework. *Comput. Chem. Eng.* **2016**, *93*, 87–100. [CrossRef]

29. Ma, J.J.; Du, G.; Zhang, Z.K.; Wang, P.X.; Xie, B.C. Life cycle analysis of energy consumption and CO_2 emissions from a typical large office building in Tianjin, China. *Build. Environ.* **2017**, *117*, 36–48. [CrossRef]

30. Skaar, C.; Labonnote, N.; Gradeci, K. From Zero Emission Buildings (ZEB) to Zero Emission Neighbourhoods (ZEN): A Mapping Review of Algorithm-Based LCA. *Sustainability* **2018**, *10*, 2405. [CrossRef]

31. EN 15978. *Sustainability of Construction Works-Assessment of Environmental Performance of Buildings-Calculation Method*; European Committee for Standardization: Brussels, Belgium, 2011.

32. ISO 14040. *Environmental Management—Life Cycle Assessment—Principles and Framework*; International Organization for Standardization: Geneva, Switzerland, 2006.

33. ISO 14044. *Environmental Management—Life Cycle Assessment—Requirements and Guidelines*; International Organization for Standardization: Geneva, Switzerland, 2006.

34. Selamawit, F.; Schlanbudch, R.; Sørnes, K.; Inman, M.; Andresen, I. *A Norwegian ZEB Definition Guideline*; The Reserch Centre on Zero Emission Buildings (ZEB), ZEB Project Report (29) SINTEF; Academic Press: Oslo, Norway, 2016.

35. Dokka, T.H.; Sartori, I.; Thyholt, M.; Lien, K.; Lindberg, K.B. *A Norwegian Zero Emission Building Definition*; Passihus Norden: Gothenburg, Sweden, 2013.

36. *Information Office of the State Council, China's Energy Policy 2012 (White Paper)*; Information Office of the State Council: Beijing, China, 2012.

37. National Development and Reform Commission of China. *13th Five-Year Plan of Railway Development*; National Development and Reform Commission of China: Beijing, China, 2017.

38. *Ministry of Housing and Urban-Rural Development of China, GB 50189-2015: Design Standard for Energy Efficiency of Public Buildings*; China Architecture and Building Press: Beijing, China, 2015.

39. National Railway Administration. *TB10100-2018 Code for Design of Railway Passenger Station*; National Railway Administration: Beijing, China, 2018.

40. Wang, J.; Liu, C. Investigation and analysis of energy-efficient potential to the space and form of high-speed railway station building. *Build. Energy Effic.* **2019**, *2*, 41–49. (In Chinese)

41. Malmqvist, T.; Nehasilova, M.; Moncaster, A.; Birgisdottir, H.; Nygaard, F. Design and construction strategies for reducing embodied impacts from buildings—Case study analysis. *Energy Build.* **2018**, *166*, 35–47. [CrossRef]

42. Yu, D.; Tan, H.; Ruan, Y. A future bamboo-structure residential building prototype in China: Life cycle assessment of energy use and carbon emission. *Energy Build.* **2011**, *43*, 2638–2646. [CrossRef]

43. Venkatarama Reddy, B.V.; Leuzinger, G.; Sreeram, V.S. Low embodied energy cement stabilised rammed earth building. *Energy Build.* **2014**, *68*, 541–546. [CrossRef]

44. Kumanayake, R.; Luo, H.; Paulusz, N. Assessment of material related embodied carbon of an office building in Sri Lanka. *Energy Build.* **2018**, *166*, 250–257. [CrossRef]

45. Seo, S.; Zelezna, J.; Birgisdottir, H. Passer, A. *Evaluation of Embodied Energy and CO2eq for Building Construction (Annex 57)*; Institute for Building Environment and Energy Conservation: Tokyo, Japan, 2016; Available online: http://www.iea-ebc.org/Data/publications/EBC_Annex_57_ST2_Literature_Review.pdf (accessed on 5 October 2019).

46. Maddalena, R.; Roberts, J.J.; Hamilton, A. Can Portland cement be replaced by low-carbon alternative materials? A study on the thermal properties and carbon emissions of innovative cements. *J. Clean. Prod.* **2018**, *186*, 933–942. [CrossRef]

47. Ministry of Housing and Urban-Rural Development of China. *GB 50352-2019. Uniform Standard for Design of Civil Buildings*; Ministry of Housing and Urban-Rural Development of China: Beijing, China, 2019.

48. Zhang, X. *Research on the Quantitative Analysis of Building Carbon Emissions and Assessment Methods for Low-carbon Buildings and Structures*; Harbin Institute of Technology: Harbin, China, 2018. (In Chinese)

49. Luo, Z. *Study on Calculation Method of Building Life Cycle CO2 Emission and Emission Reduction Strategies*; Xi'an University of Architecture and Technology: Xi'an, China, 2016. (In Chinese)

50. Roh, S.; Tae, S. Building Simplified Life Cycle CO_2 Emissions Assessment Tool (B-SCAT) to Support Low-Carbon Building Design in South Korea. *Sustainability* **2016**, *8*, 567. [CrossRef]

51. Raji, B.; Tenpierik, M.J.; Van den Dobbelsteen, A. Early-Stage Design Considerations for the Energy-Efficiency of High-Rise Office Buildings. *Sustainability* **2017**, *9*, 623. [CrossRef]

52. Ministry of Housing and Urban-Rural Development of the People's Republic of China. *GB/T 51366-2019 Standard for Building Carbon Emission Calculation*; Ministry of Housing and Urban-Rural Development of the People's Republic of China: Beijing, China, 2019.

53. Hou, P.; Wang, H.T.; Zhang, H.; Fan, C.D.; Huang, N. GreenHouse gas emission factors of Chinese power grids for organization and product carbon footprint, Zhongguo Huanjing Kexue. *China Environ. Sci.* **2012**, *32*, 961–967. (In Chinese)

54. Standard Quota Institute. *TY01-31-2015 Consumption Quota of Housing Construction and Decoration Engineering*; China Planning Press: Beijing, China, 2015. (In Chinese)

55. Zhao, M.; Künzel, H.M.; Antretter, F. Parameters influencing the energy performance of residential buildings in different Chinese climate zones. *Energy Build.* **2015**, *96*, 64–75. [CrossRef]

56. Mechri, H.E.; Capozzoli, A.; Corrado, V. USE of the ANOVA approach for sensitive building energy design. *Appl. Energy* **2010**, *87*, 3073–3083. [CrossRef]

57. Groen, E.A.; Bokkers, E.A.M.; Heijungs, R.; de Boer, I.J.M. Methods for global sensitivity analysis in life cycle assessment. *Int. J. Life Cycle Assess.* **2017**, *22*, 1125–1137. [CrossRef]

58. Budavari, Z.; Szalay, Z.; Brown, N.; Malmqvist, T.; Peuportier, B.; Zabalza, I.; Krigsvoll, G.; Wetzel, C.; Cai, X.; Staller, H.; et al. *LoRE-LCA-Deliverable 5.2 Methods and Guidelines for Sensitivity Analysis, Including Results for Analysis on Case Studies*; SINTEF: Trondheim, Norway, 2011; Available online: https://www.sintef.no/globalassets/project/lore-lca/deliverables/lore-lca-wp5-d5.2-emi_final.pdf (accessed on 20 October 2019).

59. ZEB Tool Manual. *Technical Guide. Version 1, The Research Centre for Zero Emission Buildings*; Internal Memo: Trondheim, Norway, 2017.

60. National Railway Administration. *TB/T 10429-2014: Evaluation Standard for Green Railway Stations*; National Railway Administration: Beijing, China, 2014.

61. Cheng, B.; Li, J.; Tam, V.W.Y.; Yang, M.; Chen, D. A BIM-LCA Approach for Estimating the Greenhouse Gas Emissions of Large-Scale Public Buildings: A Case Study. *Sustainability* **2020**, *12*, 685. [CrossRef]
62. Graabak, I.; Bakken, B.H.; Feilberg, N. Zero emission building and conversion factors between electricity consumption and emissions of greenhouse gases in a long term perspective. *Environ. Clin. Technol.* **2014**, *13*, 12–19. [CrossRef]

Article

Estimating Lifetimes and Stock Turnover Dynamics of Urban Residential Buildings in China

Wei Zhou [1,2,*], **Alice Moncaster** [2,3], **David M Reiner** [1] **and Peter Guthrie** [2]

[1]   Energy Policy Research Group, University of Cambridge, Cambridge CB2 1AG, UK
[2]   Department of Engineering, University of Cambridge, Cambridge CB2 1PZ, UK
[3]   School of Engineering and Innovation, Open University, Milton Keynes MK7 6AA, UK
*   Correspondence: wz282@cam.ac.uk

Received: 5 June 2019; Accepted: 5 July 2019; Published: 8 July 2019

Abstract: Building lifetime and stock turnover are both key determinants in modelling building energy and carbon. However in China, aside from anecdotal claims that urban residential buildings are generally short-lived, there are no recent official statistics, and empirical data are extremely limited. We present a system dynamics model where survival analysis is used to characterise the dynamic interplay between new construction, aging, and demolition of residential buildings in urban China. The uncertainties associated with building lifetime were represented using a Weibull distribution, whose shape and scale parameters were calibrated based on official statistics on floor area up to 2006. The calibrated Weibull lifetime distribution allowed us to estimate the dynamic stock turnover of Chinese urban residential buildings for 2007 to 2017. We find that the average lifetime of urban residential buildings was around 34 years, and the overall residential stock size reached 23.7 billion m^2 in 2017. The resultant age-specific sub-stocks provide a baseline for the overall stock, which—along with the calibrated Weibull lifetime distribution—can be used in further modelling and for analysis of policies to reduce the whole-life embodied and operational energy and CO_2 emissions in Chinese residential buildings.

Keywords: building stock; survival analysis; lifetime distribution; system dynamics

---

1. Introduction

As the largest energy consumer and CO_2 emitter worldwide [1–4], China's progress along a low-carbon development pathway has a significant impact on global efforts towards climate change mitigation. For the building sector, China is a major driving force of energy and emissions growth. According to the World Energy Outlook [5–8], final energy consumption of buildings in China increased by approximately 53% from 1990 to 2012, when China (480 million tonnes of oil equivalent, Mtoe) overtook the US (464 Mtoe), becoming the world's largest building energy consumer. By 2014, buildings in China were consuming 529 Mtoe, representing 18% of the total energy demand of buildings globally. Underlying the continued and rapid increase of building energy consumption in China are the increasing population, rapid urbanisation and consistently strong economic growth that have been driving the expansion of building floorspace and demand for energy services and thermal comfort in buildings [9]. Over the past decade, between 2.4 and 3.4 billion m^2 of new building floor area were constructed every year [10]. By 2015, the stock was estimated to have reached 57.2 billion m^2, representing 25.6% of global total building floor area (223.4 billion m^2) [11]. The expansion of building stock far outpaced the energy efficiency improvement gained from more stringent building design standards and enhanced performance of building service systems and equipment, thereby leading to continued increase in total energy consumption [12–14]. This presents a critical challenge to the Chinese government's pledge to peak its overall emissions by 2030 [15,16]. Strategically, China's building sector needs to undergo a transition towards low-carbon development whereby the increase of annual building energy consumption and emissions begins to decelerate. This challenge urgently calls for a sector-specific policy and regulatory framework governing the envisaged sector-wide transformation.

To inform policies, it is essential to have a holistic and in-depth understanding of the current status of the existing building stock, which will serve as the baseline for assessing possible future energy and carbon trajectories under various policy scenarios. This is problematic in the Chinese context not only because of the magnitude of the building stock, but also because of the lack of authoritative official statistics. The latest publicly accessible statistics on urban residential building stock were published in the China Statistical Yearbook in 2006. Therefore, forecasting possible trajectories of building stock expansion and the associated energy and emissions for the future starting from 2019 will first and foremost require estimating how the building stock has evolved over the period from 2006 to 2018. The evolution and expansion of the building stock is driven by the dynamic interplay between new construction, meeting incremental demand growth as a result of economic growth and rising living standards, existing buildings remaining in use but undergoing an ageing process, and old buildings, which are either physically demolished or functionally disused.

Building lifetime is a critical factor in the dynamic relationship between old and new buildings in the stock. In the Chinese context, it has been suggested that the average lifetime of buildings is as short as 25–30 years in urban areas and 15 years or even less in rural areas [17–20]—significantly shorter than buildings in developed countries [21,22]. The short lifetimes are due to various factors, including quality of building materials, design standards, construction techniques and practices, maintenance and renovation, inappropriately accelerated demolition as a result of rapid urbanisation and city rebuilding, etc. [23,24].

The high "turnover" rate reflects the fact that the building stock is being constantly rapidly replenished as a result of old buildings with short lifetimes being removed and new buildings being constructed to meet demands, as well as the great complexity and uncertainty associated with building-stock characteristics. These have significant implications for stock-wide energy use and emissions over the medium- to long-term. On the one hand, a faster turnover rate would imply that the Chinese residential building stock is less prone to the risk of operational energy and carbon "lock-in" [22,25–28] compared to its counterparts in other countries with longer lifetimes and thus slower turnover rates. On the other hand, however, turnover has significant implications for building energy from a lifecycle perspective. Massive construction and demolition require significant amounts of energy for building materials production, construction activities, demolition and disposal, collectively known as "embodied energy" [29]. From 2004 to 2012, the embodied energy of buildings in China steadily increased [30], reaching 455 Mtoe in 2013, and accounting for 16% of China's total primary energy consumption [14]. This is comparable in size to the 529 Mtoe used in operational energy in existing buildings in 2014. The considerable impact of embodied energy and carbon due to building stock turnover must therefore not be overlooked in the decision-making process for sustainable design, construction and use of buildings in order to make real reductions in greenhouse gas (GHG) emissions [31].

The two conflicting arguments suggest the need for some sort of trade-off, but both clearly demonstrate that building stock turnover is key to understanding total life cycle energy impacts from buildings. Building lifetime is a determinant factor underlying the dynamics of building stock turnover. We therefore present a system dynamics model to calibrate residential building lifetime in China and estimate the stock size and age profile for the recent historical period of 2007 to 2017.

The rest of the paper is organised as follows. Section 2 presents a review of literature closely relating to Chinese building stock, identifies some common issues associated with the methodological approaches taken and justifies the relevance of the present study. Section 3 introduces the methodological framework, which is followed by Section 4 dedicated to conceptualising and developing the model structure and discussing the empirical data needed to parameterise the model. In Section 5, the modelling results are presented and compared with other studies. Section 6 discusses potential applications of the model to the whole-life energy and carbon of buildings and draws some conclusions.

2. Literature Review

Whilst the building stock has a fundamental impact on macro-level energy consumption and carbon emissions with significant policy implications, studies investigating the complexities and dynamics of building stock appear to be rather limited—substantially less than the wide range of studies at the individual building level.

One of the first simplified models of the Chinese building and infrastructure stock was developed by Yang and Kohler [23], who set 2005 as the base year. The existing stock in 2005 was set as the initial stock, which included buildings built between 1978 and 2005, and was assumed to be composed of several age cohorts. For buildings built from 2005 onwards, a cohort-based approach was applied to define the average age of buildings to model the stock evolution during the period from 2005 to 2050 on a five-year basis. While their model did not explicitly represent the dynamically aging process of buildings, Yang and Kohler [23] pointed out the considerable influence of the probable lifespan of buildings on future mass flows and environmental impacts. Similar to Yang and Kohler [23], a static approach was taken by the High Efficiency Buildings model developed by the Centre for Climate Change and Sustainable Energy Policy, known as the 3CSEP-HEB model [32]. Their model assumed the Chinese building stock had an annual demolition rate of 0.5% and a retrofit rate of 1.4%, which implied a homogeneity of Chinese buildings in terms of lifetime.

Taking a more dynamic perspective, Hu, Bergsdal, et al. [18] analysed the Chinese building stock by assuming a normal distribution function for building's lifetime. The size of building stock was estimated using population and per-capita floor area sourced from China Statistical Yearbooks. The authors explored various scenarios of future demand for building materials such as steel and concrete for Chinese residential buildings, both at the national level [33] and at the city level [34]. In fact, the study by Hu, Bergsdal, et al. [18] appears to have been the first attempt to apply the concept of lifetime distribution to the Chinese building stock. For example, drawing from the work by Hu, Bergsdal, et al. [18], Huang et al. [35] carried out a similar study investigating the materials demand and environmental impact of buildings, where it was assumed that the lifetime of concrete buildings is normally distributed, assuming an average lifetime of 30 years for brick-concrete buildings and 40 years for reinforced concrete buildings. In Hong et al. [36], the same methodological approach was applied to develop a building stock turnover model underlying the projected trajectories of demand for building materials and the corresponding embodied energy over the period from 2010 to 2050. Both residential and commercial buildings were assumed to follow normal distributions in terms of their lifetimes, with the standard deviation being set to be 1/3 of the average lifetime. Similarly, investigating the impact of technical progress and the use of renewable energy in the building sector over the period from 2010 to 2050, Shi et al. [37] applied the China TIMES model and represented the lifetime distribution of buildings using a normal distribution. In these studies, historical per-capita floor area was sourced from China Statistical Yearbooks to derive possible future trajectories of per-capita floor area and overall floor area.

It is noteworthy that the use of historical per-capita floor area published in past China Statistical Yearbooks used in building stock and energy modelling has also been common in previous studies that were less explicit or dynamic, often where buildings were one component of a much larger model. For example, the residential building sector in the China End-Use Energy Model developed by the Lawrence Berkeley National Laboratory used per-capita floor area from the Yearbooks to estimate the overall stock size and develop a 2050 outlook for energy and emissions [38–41]. Yu et al. [42] studied the potential impacts of alternative building energy code scenarios on building energy use and associated emissions in China, using a detailed building energy model nested in the Global Change Assessment Model (GCAM) long-term integrated assessment framework [43]. In their model, Yearbook data on per-capita floor area were used to calibrate the modelling of floor area expansion before 2010, and it was assumed that buildings would retire at an annual rate of 1/30 of the remaining stock. Delmastro et al. [44] developed the Energy for Buildings (EfB) model as part of the Energy Demand Projection Model for China (EDPM-CN) to analyse the residential energy consumption trends up to

2030 under various technological and policy scenarios. Along with other drivers of energy demand in their model, the historical per-capita floor area published in the Yearbooks was a key variable. Similarly, to explore various scenarios of carbon emissions from Chinese buildings through 2050, Yang et al. [45] applied a grey modelling technique based on the historical per-capita floor area data published in past China Statistical Yearbooks to forecast the future trend of building stock size for the period up to 2020 and scenario analysis for the period beyond 2020.

Essentially, there are three main methodological concerns associated with previous research: (i) arbitrary choice of mean and standard deviation; (ii) ambiguity associated with existing building stock size and age profile in the start year for the modelling; and (iii) use of per-capita floor area data, leading to inflated estimates. Firstly, the few studies using a distribution to represent the lifetime of buildings assumed that building lifetime is normally distributed. For a substantially under-researched area, as discussed below in Section 4.2, there will inevitably be multiple approaches that could be taken, and a normal distribution is not unreasonable. However, what is more critical and potentially questionable is the approach of defining the parameters of the normal distribution without validation or calibration using empirical data. In previous studies, the mean representing the average building lifetime was assumed to take values in the range of 30–50 years. Furthermore, the standard deviation was commonly assumed to be 30% of the mean. The range of values for the mean was based on anecdotal evidence drawn from limited or individual cases. Moreover, the assumption used for the standard deviation was purely arbitrary. Taken together, these parameters specify inadequately substantiated shapes of the distributions that govern the lifecycle of buildings, thereby rendering high uncertainties of the building stock turnover and calling into question the calculations of energy consumption and carbon emissions that build on the stock turnover dynamics.

In addition to the somewhat arbitrary parameters used in the distributions, there is a general absence of detail on how the initial stock in the start year of the observation period is treated in the models. Regardless of which specific year is chosen as the start year for modelling, the size of the existing stock in that year and especially the age profile of the buildings are key determinants of the subsequent evolution of the stock. For a given stock size with a pre-defined normal distribution governing the lifetime distribution (e.g., a mean value of 50 and a standard deviation of 15), the age profile of the existing buildings in the stock makes a significant difference in the remaining lifetime of buildings and the overall size of the stock. As an extreme example, an age profile with over 90% of buildings younger than 30 years means the stock will remain standing much longer than an age profile with over 90% of buildings older than 70 years. Moreover, the removal of buildings from the stock to a large extent determines the incoming new buildings which will then be subject to various probabilities of demolition over time. Hence, the age profile in the initial stock technically has a knock-on effect on stock evolution over time. Of course, such effects may be marginal for a very small initial stock in a start year that is several decades ago if the modelling focus is on current status or future trends. However, for an initial stock in a fairly recent year (e.g., 2010 in the case of [37] and [36]), its overall size is large and comparable with the stock size for the modelling period. The composition of the stock in terms of building age profiles cannot be overlooked. In fact, the simplified model by Yang and Kohler [23] was the only one we could find which explicitly but briefly introduced how the initial stock in the base year was treated in the modelling. However, the lack of sufficient quantitative detail makes it difficult to gain a full understanding of their considerations and therefore evaluate the implications for modelling results relating to not only stock itself but also energy. In summary, a review of these studies suggests that there has been a general inadequacy of robustness and transparency with regard to applying a normal distribution to building stock.

Finally, directly using the per-capita floor area for urban residential buildings as published in the China Statistical Yearbooks leads to substantially inflated estimates of building stock size. The annual data reported in the yearbooks, supplied by the Ministry of Housing and Urban-Rural Development (MOHURD), was collected through sampling targeting urban family households with a registered permanent residence. The sampling excluded urban dwellers without permanent residence status, such as university students and young professionals recently graduated from universities who usually have so-called "collectively registered" status, as well as a large number of unregistered rural migrant workers living in cities. As opposed to the registered family households, they are not "permanent" but "floating" in that most are not homeowners and instead rent their accommodations. The per-capita floor area of the floating population is substantially less than that of registered family households, and their accommodation conditions were not reflected in the sampling [46–48]. This means that the Yearbook data on per-capita floor area for urban residential buildings has been over-estimated [49,50]. Accordingly, multiplying the over-estimated per capita data by actual urban population data to derive the total stock size will in no way reflect the real situation. In fact, the real situation in terms of total stock size of urban residential buildings in China is unknown, at least for the most recent historical period of 2007 to 2018 (MOHURD stopped publishing official statistics after 2006). The absence of official statistics makes calibration of model estimates impossible, and therefore at least partially explains the considerable variation in estimated building stock size over this period as found in the literature. For example, estimates of the urban residential building stock size in 2010 ranged from 14 billion m^2 [45] to 17 billion m^2 [35], 20 billion m^2 [18,23], and 21 billion m^2 [36]. To obtain a reasonably accurate estimate of the overall stock size of urban residential buildings, either the Yearbook data should be adjusted downwards by accounting for the urban population who rent, or some alternative method of avoiding the use of per-capita floor area.

The problems discussed above suggest a research gap that needs to be addressed. It is not meant to downplay the value of previous studies which collectively made significant contributions to this under-researched but fundamentally and strategically important area. Inspired by and building on previous studies, we propose an alternative method to estimate the Chinese urban residential building stock. Underlying the estimated stock is a calibrated building lifetime distribution function, which can serve as the basis for further modelling and policy analysis on building lifetime energy consumption and carbon emissions.

3. Methodology

We used system dynamics to model the stock turnover of urban residential buildings in China. System Dynamics is a modelling paradigm focusing on dynamic complexity arising from the structure, feedbacks, non-linearity and time lags of the system in question [51–53]. The model was developed and implemented using Vensim, a commercial software for System Dynamics modelling [54], as well as using the R statistical computing and graphics environment [55]. The model presented here treats building stock evolution as a continuous process of introducing new cohorts which age over time, capturing the dynamic interplay between new construction, operation and demolition. Building lifecycle is regarded as a survival process subject to various factors. Demolition of buildings is modelled as a stochastic process based on a hazard function derived from a Weibull distribution, whose parameters were estimated using historical data relating to building stock.

4. Model Development

4.1. Dynamics in Building Demolition

A fundamental consideration of dynamic building stock is that future development is strongly influenced by past activities. In particular, building demolition activities are likely to be a function of construction activities in previous years and the expected lifetimes of buildings in use. By the end of a year, the total volume of demolition will be the sum of all existing buildings constructed in previous years that have reached the end of their lifetimes, for whatever reasons, in this particular year. Similarly, the buildings remaining in the stock are those which are either newly constructed in this particular year or those which were previously constructed but have not reached the end of their lifetime. We acknowledge that a building may be disused functionally but still not demolished physically. Since the ultimate interest of modelling building stock turnover is in energy consumed by buildings, a functionally disused building no longer consumes energy and therefore was considered equivalent to a physically demolished building from an energy perspective. Hence, in the rest of this study, demolition and disuse are used interchangeably.

There could be various factors which accelerate demolition. While the degrees to which different factors play out are context specific and therefore may differ significantly, the explicit direct result is a fast "turnover" of building stock. Therefore, building lifetime is critical to the turnover dynamics of building stock. There is often a lack of authoritative statistics relating to building lifetime, particularly in developing countries. At the country level, given the huge volume of buildings and significant heterogeneity in terms of their physical characteristics and socio-economic contexts, it would be highly unrealistic to expect that buildings constructed and put into use in various cities across a country in a given year would be in service for exactly the same period and then demolished/disused simultaneously. Hence, it would be inappropriate to use a constant to represent building lifetime. As an intuitive and reasonable alternative, a profile in some form of probability density function (PDF) can be used to approximate the likely lifetime distribution of buildings constructed in a given year, so as to recognise and represent the uncertainties associated with the factors collectively influencing lifetime of buildings.

Mathematically, for any year t, the amount of demolition is the integral of the new construction in year s weighted by the probability of the new construction in year s having a lifetime equal to $(t - s)$ years. That is to say, those buildings constructed in year s with a lifetime of $(t - s)$ years will have reached the end of their lifetime and therefore will have been demolished or disused in year t. Interpreted in another way, the curve expressed by (1 − the cumulative distribution function (CDF) of the lifetime profile) represents the probability that a building constructed and put into use in year s remains in use in year t.

$$Demolition(t) \;=\; \int_{t_0}^{t} p(t-s) Construction(s) ds \tag{1}$$

In year s, the building stock accumulates the difference between new construction as input and demolition as output. The accumulated difference is the buildings that remain in use in year t. In the most simplistic situation where the stock is initially empty and new buildings are constructed in base year t_0 only and not in any following years, the stock in year t is a mix of buildings belonging to this same cohort, which is the group of buildings built in the same year, but with different remaining lifetimes. If new buildings are built every year, then the stock in year t has $(t - t_0 + 1)$ cohorts of buildings. In a more general situation where new buildings are constructed every year and the initial stock includes existing buildings constructed previously (e.g., for each of the past 10 years before year t_0), then the stock in year t has $(t - t_0 + 1 + 10)$ cohorts of buildings.

4.2. Building Survival Analysis

The dynamic lifecycle of buildings suggests the suitability of applying the general framework and key features of survival analysis to this research. Survival is used to describe a lifespan or a living process involving sequential occurrences of events or status changes. Survival analysis emphasises describing, measuring and analysing the events of interest for making predictions about not only survival itself but also the so-called "time-to-event", which refers to the length of time until the occurrence of an event or the change of status [56,57].

As this research targets the Chinese residential building stock, which is on a scale of billions of square meters, it is worth clarifying that a "building" in this context is taken as a flexible and continuous aggregation of floor area (measured in square meters) that can be "partially" demolished or disused, mathematically. The physical demolition or functional disuse of a building designates the particular "event" of interest, whose time at occurrence is the "time-to-event" in the survival analysis framework. At a given time, for a building in use, a predicted time in the future at which the building will be physically demolished or functionally disused means the termination of the survival process of the building—namely, the end of the building's life. The corresponding time interval is the expected remaining lifetime of the building. If the time at which the building was originally constructed and put into use is taken as the beginning of the time interval, then the time interval is the expected entire lifetime of the building. In this context, the expiry of building lifetime due to physical demolition or functional disuse is modelled as a stochastic process based on a hazard function. Conceptually, the hazard function represents the conditional probability that a building will expire in year $t + 1$, provided that it has successfully survived to year t. Mathematically, the hazard function is the ratio of the lifetime PDF to the survival function, which is the complement of lifetime CDF.

In general, a range of parametric survival distribution functions are available to describe the survival process in various fields. However, there is extremely limited literature on survival analysis or lifetime data analysis of buildings. In probably one of the most relevant studies, Miatto et al. [58] tested various PDFs and found that lognormal distribution offered the best fit to a large volume of real data on the lifespans of buildings in Nagoya and Wakayama, Japan, where buildings were short-lived, with average lifespans shorter than 30 years. The same study also conducted a smaller-scale study on buildings in Salford, UK, where buildings were much older, and a Gompertz distribution turned out to best fit the lifespan data. They pointed out that the lack of a proper building cohort dataset was a fundamental issue facing building material stock accounting, such as building demolition wastes. From an economic perspective, buildings can be regarded as a type of capital asset, hence building stock can be regarded as capital stock [59,60]. A range of PDFs have been used as proxies to approximately represent service lives and retirement/discard patterns of capital stocks in different countries, including normal, log-normal, Gompertz, Weibull, and Winfrey distributions [59–64]. For example, in the Dutch survey, information on capital stocks and capital discards was used to compute empirical survival probabilities of various types of asset and industry—including buildings, which were found to be well approximated by two-parameter Weibull distributions [60].

We used the Weibull distribution to approximate the lifetime distribution of urban residential buildings in China. The Weibull distribution has been widely used as the functional form for lifetime distribution in mortality and reliability applications [60,65,66]. It is defined through a shape parameter α ($\alpha > 0$) and a scale parameter λ ($\lambda > 0$). Table 1 summarises the key representations of the Weibull distribution. Specifying any one of the four representations allows the other three to be ascertained.

Table 1. Weibull distribution. CDF: cumulative distribution function; PDF: probability density function.

PDF	$f(x) = \left(\dfrac{\alpha x^{\alpha-1}}{\lambda^{\alpha}}\right)e^{\left(-\left(\frac{x}{\lambda}\right)^{\alpha}\right)}$
CDF	$F(x) = 1 - e^{\left(-\left(\frac{x}{\lambda}\right)^{\alpha}\right)}$
Survival function	$R(x) = e^{\left(-\left(\frac{x}{\lambda}\right)^{\alpha}\right)}$
Hazard function	$h(x) = \dfrac{\alpha}{\lambda}\left(\dfrac{x}{\lambda}\right)^{\alpha-1}$

In the above table, if x is used to represent "time-to-failure", the Weibull distribution is characterised by the fact that the hazard function is proportional to a power of time. Hence, the shape parameter α can be interpreted as a measure of change in the risk of an asset (e.g., a building, failing and therefore being demolished/disused). For $0 < \alpha < 1$, the risk decreases over time, suggesting significant "infant mortality". For $\alpha = 1$, the risk remains constant throughout an asset's lifetime. For $\alpha > 1$, the Weibull distribution exhibits characteristics of decay, where the older the asset is, the more likely it will fail in the near future. This indicates an "aging" process in which the value of α determines the shape of the hazard function curve (i.e., the change of failure rate over time) [57,60,65,67–69]. In practice, it would be more reasonable to expect buildings to have an α above 1 than below 1.

4.3. Model Structure and Components

The shape and scale parameters of the Weibull distribution were calibrated so that the approximated building lifetime distribution would fit the empirically observed data to the greatest extent possible.

However, there are no official statistics on building ages in China, and past studies related to building lifetime in China are limited, as discussed above. Therefore, the calibration was performed by achieving the best-possible fit of the modelled annual aggregated floor area to the corresponding historical data on floor area, using the emergent behaviour of the building stock turnover model. To do so, the building stock in the turnover model was disaggregated into a series of cascading sub-stocks of buildings forming an "aging chain", with each sub-stock representing a particular building age group. The basic mechanism is that, on the aging chain, sub-stock j receives the outflow of sub-stock $j - 1$ as its inflow, undergoes an aging process, and subsequently sends its outflow to sub-stock $j + 1$ (Figure 1).

Figure 1. Basic mechanism of a simple aging chain.

The removal of demolished/disused buildings from each sub-stock is subject to its age-specific hazard rate, which is determined by the shape and scale parameters of a Weibull distribution. The age group duration represents the length of time that buildings in use reside in a sub-stock before shifting to the next sub-stock in the chain (Figure 2). With age group duration set to 1 year, the chronological aging process is discretised (i.e., each sub-stock represents buildings within a one-year age group—for instance, 30-year-old buildings are in a different sub-stock than 31-year-old buildings). This level of granularity offers a detailed representation of sub-stocks characterised by heterogeneity with respect to age (and energy-related properties, provided that additional layers are added to the model). In so doing, the aging process of buildings can be separately tracked and allow experimentation with policy interventions targeting buildings of specific age groups.

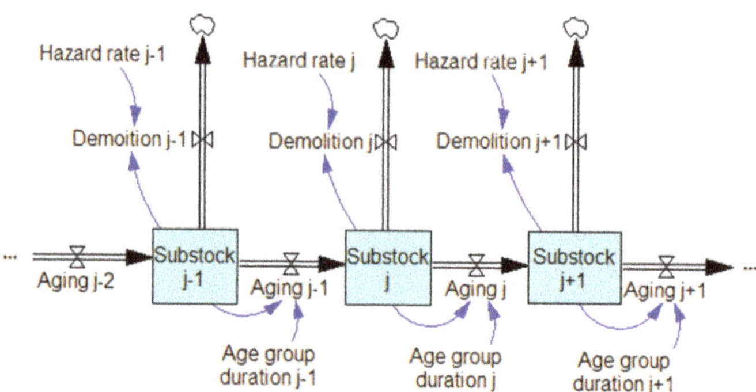

Figure 2. Aging chain with explicit modelling of sub-stock specific demolition.

Given that building lifetime in China is generally short, the model uses 101 sub-stocks to respectively represent 0-year-old buildings, 1-year-old buildings, 2-year-old buildings, and so on. The 101st sub-stock represents all buildings that are 100 years old or older, but such buildings account for a negligible percentage of the overall stock of urban buildings in China. Indeed, buildings aged below 40 years accounted for more than 95% in the total stock of urban residential buildings in 2010 [70]. The schematics in Figures 1 and 2 were converted to the working model using subscripting techniques. The model structure was re-formulated to improve representation and analytical convenience. As shown in Figure 3, the building stock can be viewed as a stack of 101 sub-stocks. For any given year, the total stock of buildings in use is the sum of the age-specific building sub-stocks, and the total demolition of buildings is the sum of the age-specific demolitions.

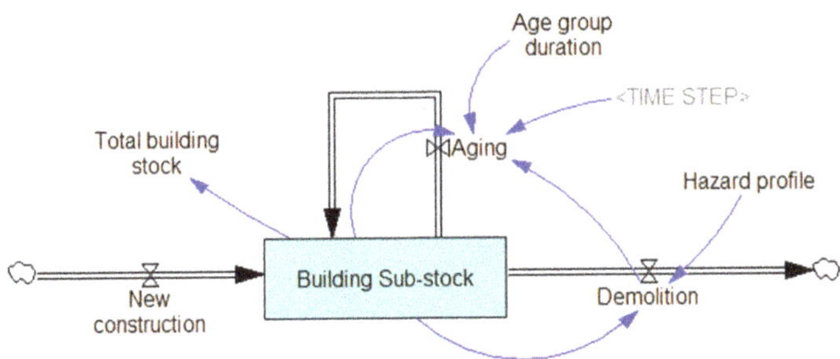

Figure 3. Aging chain structure in the model.

Subscripts are applied to each sub-stock, to the aging and demolition rates, and to the hazard profile. The hazard profile consists of age-specific hazard rates calculated from the hazard function. The numbering of subscripts represents the age of buildings. For example, Building Sub-stock[10] represents the group of 10-year-old buildings, Demolition[10] represents the demolition and removal of buildings from Building Sub-stock[10], whereas Aging[10] represents the shift of existing buildings from Building Sub-stock[10] to Building Sub-stock[11]. Mathematically, the relationships between these variables are expressed as follows:

$$Aging[j] = \frac{Building\ Substock[j] - Demolition[j] * Time\ Step}{Age\ group\ duration}, \tag{2}$$

$$Demolition[j] = Building\ Substock[j] * Hazard\ profile[j], \tag{3}$$

$$Building\ Substock[j] = \begin{cases} New\ construction - Demolition[j] - Aging[j], & for\ j = 0 \\ Aging[j-1] - Aging[j] - Demolition[j], & for\ j = 1, 2, \dots 100. \end{cases} \tag{4}$$

With the age group duration and time step both set to be equal to 1 year, the aging equation can be re-written as follows.

$$\begin{aligned} Aging[j] &= \frac{Building\ Substock[j] - Demolition[j] * Time\ Step}{Age\ group\ duration} \\ &= \frac{Building\ Substock[j]}{Age\ group\ duration} - Demolition[j] \\ &= Building\ Substock[j] - Demolition[j]. \end{aligned} \tag{5}$$

This simplified equation, together with the previous equations for building sub-stocks, clearly show that the aging process is effectively the discrete shift of a group of buildings from age j to age $j + 1$, duly taking into account time-varying annual demolition applicable to age j.

Figure 4 illustrates the dynamics of the building aging process. In any year, the entire stock is composed of 101 building age groups. The flow of buildings through the age groups forms the aging chain. Year by year, a building gets older and thus "relocates" itself between age groups along the chain, on the condition that it remains alive. For example, the highlighted blocks show a particular cohort of buildings constructed in 2019 moving up through the age groups/sub-stocks. Initially, the cohort stays in age group 0 in 2019, and then moves to age group 1 in 2020, age group 2 in 2021, age group 3 in 2022, and so on. Every year, new buildings are constructed, put in use, and therefore become part of the total stock, although the rates vary subject to demand driven by socio-economic development and rates at which old buildings are demolished/disused and consequently removed from the stock. Note that demolition from each of the 101 age groups is not explicitly shown in order to avoid visual clutter.

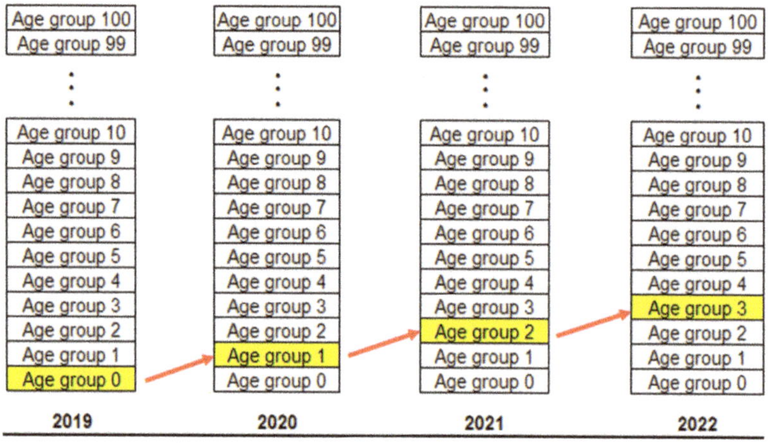

Figure 4. Dynamics of the building aging process.

4.4. Data Sources

Historical data on total floor area of building stock from authoritative sources are limited. The latest official statistics on total floor area of urban residential buildings across China was 11.29 billion m^2 for 2006, published in the China Statistical Yearbook 2007. The source of data was MOHURD. From 2007 onwards, these statistics were no longer published. Prior to 2006, data was provided in the annual China Statistical Yearbooks and in the Statistical Communique on Urban Housing released by MOHURD from 2002 to 2005. The earliest year for which data are available is 1978. Data on annual new construction of urban residential buildings was taken from China Statistical Yearbooks.

The calibration of the shape and scale parameters of the Weibull distribution for building lifetime was realised by comparing and minimising the difference between the estimated annual total floor area of urban residential buildings, which is the emergent behaviour of the building stock resulting from the dynamic process of new construction, aging and age-specific demolition, with the official statistical data over the historical period through 2006. This method allows parameters governing building lifetime to be calibrated rather than arbitrarily defined, thereby avoiding relying on the over-estimated per-capita floor area. Accordingly, the stock turnover dynamics and the resultant annual total floor area over the period of 2007 to 2017 were estimated using the officially published annual new construction data over that period and the calibrated Weibull distribution.

5. Results

Using the historical data and stock turnover model, the Weibull distribution was calibrated to have a shape parameter of 1.47 and a scale parameter of 37.64. Its PDF, CDF and hazard function are shown in Figure 5. The resultant building lifetime distribution was found to have a mean value of 34.1 years and a standard deviation of 23.5. This confirms the general observation that urban residential buildings in China have an average lifetime much shorter than the design lifetime of 50 years. The average lifetime estimate of 34.1 years is consistent with the assumptions made by previous studies that employed a normal distribution, which commonly set the average lifetime in the range of 30 to 50 years.

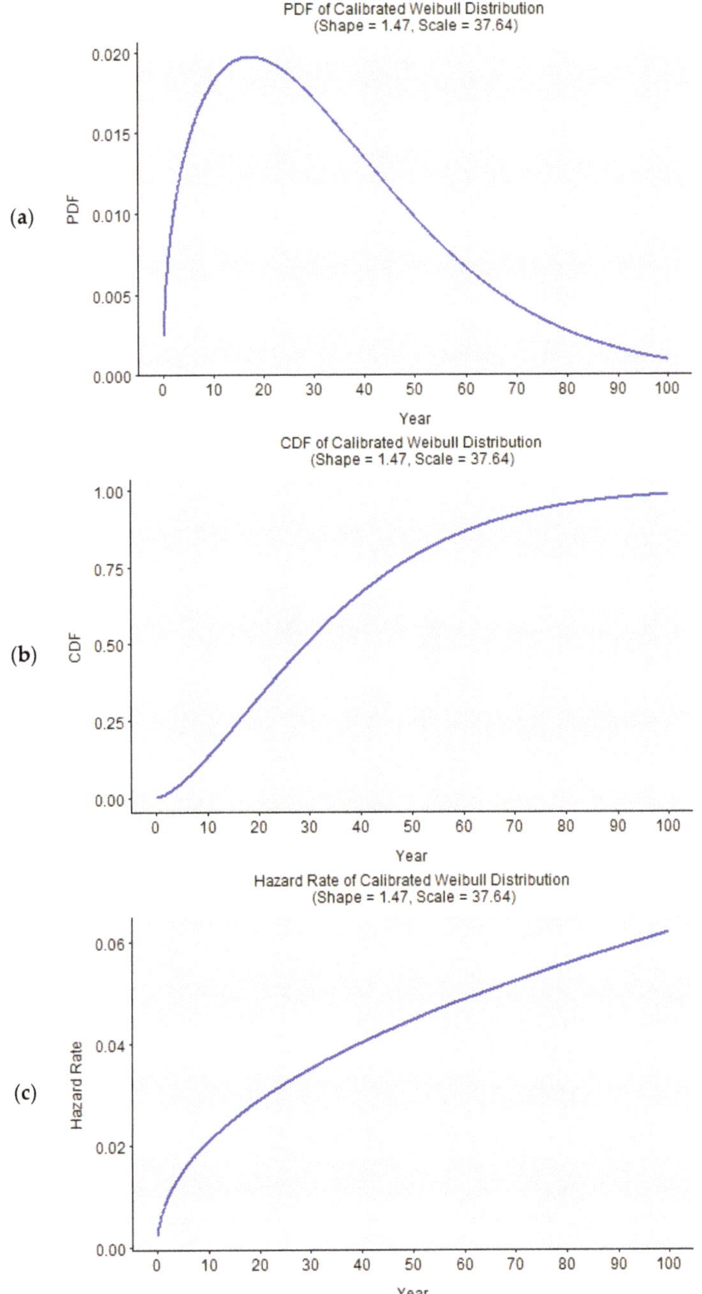

Figure 5. Weibull distributions calibrated using historical data: (**a**) PDF; (**b**) CDF; (**c**) Hazard Rate.

Due to the lack of official statistics on the annual demolition of residential buildings, it was not possible to directly cross-validate the modelling result using historical annual demolition data. In an indirect way, however, comparing the demolition estimated in our study with previous studies provides an alternative basis for evaluating the robustness of the modelling approach and the calibrated building lifetime. According to THUBERC [71], the ratio of aggregated demolished buildings to aggregated newly constructed buildings over China's 11th Five-Year Plan Period (2006 to 2010) was approximately 34%. In our study, using the calibrated Weibull parameters in the stock turnover model, this ratio was found to be 32%—very close to the THUBERC [71] estimate. In absolute terms, annual demolition level we estimate is of the same order of magnitude as previous studies. For example, for 2010, the annual demolition was estimated by this study to be 1.49 billion m^2, approximately 1.3 billion m^2 by [18] and approximately 1.7 billion m^2 by [35]. The difference was due primarily to different settings of lifetime distribution parameters.

Our modelling results suggest a continuously increasing trend of the overall stock size of urban residential buildings, increasing by 33.1% over 8 years from 17.8 billion m^2 in 2010 to 23.7 billion m^2 in 2017. Figure 6 compares the stock size of residential buildings in this study and that of the Annual Report on China Building Energy Efficiency [72] for 2010–2017. The Annual Report was developed by the leading research institution, the Tsinghua University Building Energy Research Centre (THUBERC), as part of a larger consultancy project funded by the Chinese Academy of Engineering, and is widely recognised as an authoritative report on building energy in China. As shown in Figure 6, the differences were consistently marginal, supporting the validity of the assumptions that were fed into our model.

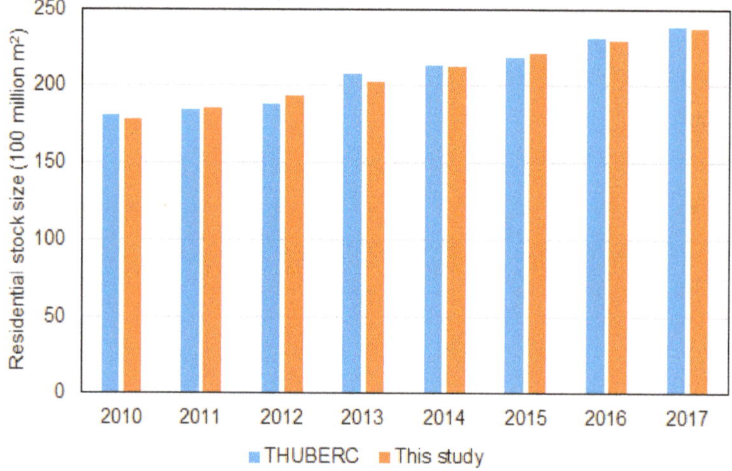

Figure 6. Comparison of residential stock size between this study and Tsinghua University Building Energy Research Centre (THUBERC) [72].

However, compared to the Annual Report, which only gives the total stock size, our model offers additional insights into the residential buildings in the form of the explicitly modelled building aging process. The total stock of residential buildings in each year is disaggregated into age-specific sub-stocks, each of which goes through an aging process subject to age-specific demolition probability estimated by the Weibull lifetime distribution (Figure 7). For each year, new buildings constructed in that year and existing buildings that remain in use in that year are spread across the age range, collectively creating the age profile of all buildings in that year. For subsequent year, the age profile changes due to new construction, aging and demolition. These on-going dynamics, which result in the turnover of the overall residential stock, are fully captured in our model. In addition, whereas the now-outdated Annual Reports provided an overview up to that given year (2006 or earlier), the Weibull lifetime distribution in our model serves as an avenue through which possible future stock turnover trajectories can be modelled. For example, given the total size and age profile of the 2017 stock, the 2018 stock can be estimated using the new construction in 2018 and the removal of existing buildings based on their respective age-specific demolition probabilities. The same logic applies going forward, where the interaction between new construction and demolition and trends in per-capita floor area will drive the growth dynamics of the stock and allow for reasonable forecasts of future growth.

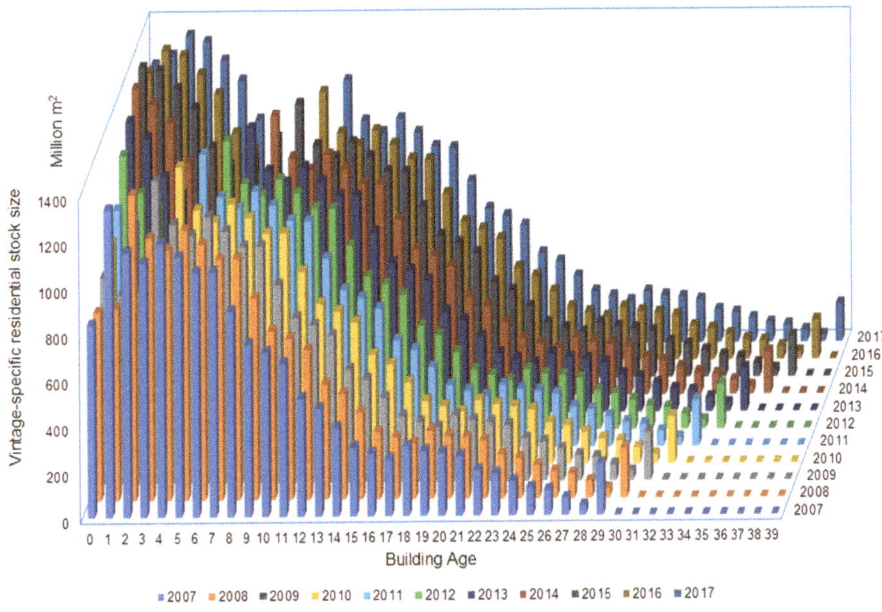

Figure 7. Aging process of age-specific residential building sub-stocks.

As shown in Figure 7, for the building age profile in each year, the oldest sub-stock of buildings is greater than the immediately preceding younger sub-stocks. For example, for 2007, the column representing the oldest sub-stock (i.e., 29 years old) is much taller than the several preceding ones, which exhibit a generally descending trend due to aging and demolition. This is because the 29-year-old sub-stock in 2007 represents all those buildings already existing in 1978 (the first year of the model) that survived to 2007. That sub-stock includes all existing buildings of various ages in 1978. Likewise, the 30-year-old sub-stock in 2008 also represents older buildings already in use in 1978 and remaining in use in 2008. So, the *x*-axis in Figure 7 is the building age for all buildings other than for this particular set (i.e., buildings in existence in 1978). It was not possible to differentiate those old buildings by their ages in the initial stock in 1978 because no statistics are available for the age profile before 1978. Therefore, without a known composition of buildings by age, the 1978 stock was treated in the model as a mixed one, in which buildings were assumed to be subject to the same probability of demolition. Mathematically, this means these buildings' lifetimes were assumed to follow an exponential distribution whose hazard function was a constant equal to the reciprocal of their average lifetime, which was taken into account in the calibration process. We acknowledge that this is a minor methodological limitation due to data availability constraints, but given its small overall size, the impact of the age profile of buildings in the initial stock in 1978 on more recent stock is largely negligible. Over time, this initial stock will only keep shrinking due to demolition, and its impact will further diminish accordingly.

6. Discussion and Conclusions

While it is generally believed that Chinese buildings are short-lived, there is a dearth of official statistics and empirical data on building lifetimes to substantiate this observation. Moreover, official statistics on total floor area of urban residential buildings only exist up to 2006, resulting in an unknown historical growth trajectory of urban residential building stock in China from 2007 onwards. Previous studies estimating recent and future Chinese building stock and energy use make various assumptions about building lifetime and stock turnover as part of their methodological approaches in an effort to overcome the lack of available data. However, as discussed earlier, these models use various questionable assumptions, as evidenced by the wide variation in results of different models. Moreover, the dynamic profile of age-specific sub-stocks and the implications for energy and carbon have not been extensively explored, which suggests a research gap in estimating building lifetime and stock turnover dynamics.

Our study developed a residential stock turnover model using a system dynamics approach. The model applied survival analysis to represent building lifecycle, from being newly constructed to being eventually demolished. Demolition was modelled as a stochastic process based on a hazard function derived from a Weibull distribution representing the uncertainties associated with building lifetime. Using historical data starting with 1978, the first year official building stock statistics are available, the Weibull distribution's shape and scale parameters were calibrated. The specified Weibull distribution had a mean value of 34.1, which represents the average building lifetime. Our result substantiates the general observation that urban residential buildings in China have an average lifetime much shorter than the design lifetime of 50 years. In the absence of official statistics on building lifetime, our study can assist policy-makers in characterising existing buildings at the national, provincial and municipal levels. Based on the calibrated lifetime distribution, the total stock size of urban residential buildings was estimated to have increased from 17.8 billion m^2 in 2010 to 23.7 billion m^2 in 2017, based on a dynamically changing age profile.

However, the value of estimating the building lifetime distribution and obtaining an explicit set of age-specific sub-stocks goes beyond understanding the dynamics of the residential building stock itself and offers three sets of insights. Firstly, the estimated lifetime distribution makes it possible to explicitly estimate annual new construction and demolition, which can be directly used to quantify the total initial and demolition embodied energy and carbon incurred every year. The impact of potentially changing the lifetime distribution on this embodied energy and carbon, via planning policy or as a result of economic and environmental factors, can be examined.

Secondly, model granularity at the level of age-specific building sub-stocks offers a detailed representation of the building stocks' heterogeneity with respect to operational energy performance. Going forward, it is reasonable to expect that new buildings will be built to higher standards of operational energy performance due to increasingly stringent design codes and technological advances. Separately tracking the aging process of different cohorts of buildings enables policy-makers to appreciate the dynamics of the stock composition of buildings with different operational performance and evaluate the trajectories of stock-wide average operational energy intensity per square metre. Explicitly modelling the aging process also provides analytical convenience to enable detailed policy experimentation, such as by targeting the retrofitting of old buildings at different ages for different depths of energy performance improvement, which makes not only technical but also economic sense.

Thirdly, and perhaps most importantly, the ability to model the temporal dynamics of the building stock enables the integration of embodied and operational impacts. A dynamic model allows us to explore their relative importance in the context of further developments in green building materials, strengthening design codes for new buildings and scaling up the energy-related retrofits of existing buildings. In so doing, a fuller understanding of lifecycle energy and carbon of urban residential buildings in China can be reached so as to better inform policies aiming to decarbonise buildings.

The whole-life (embodied plus operational) energy and carbon of the Chinese building stock are major contributors to global climate change. In 2013 and 2014, the operational energy used in existing buildings was similar in size to the embodied energy used in the construction of new buildings, together accounting for around 34% of China's total primary energy consumption. In order to reach economy-wide peak emissions by 2030, it is critical that policies effectively address both embodied and operational energy of the building stock.

We have demonstrated that the dynamics of building lifetime and building stock turnover are fundamental prerequisites to forecasting whole life energy and carbon impacts and to understanding the interactions between the two. To address the current lack of data, we have developed a calibrated model using a series of 101 heterogeneous building stocks. The model outputs indicate that the average lifetime of urban residential buildings in China is currently 34.1 years, and that the total stock has increased by 33.1%, from 17.8 billion m^2 in 2010 to 23.7 billion m^2 in 2017.

The next step will be to extend this model by adding additional properties such as embodied energy intensity and operational energy intensity, as well as additional structures and components that drive stock development, such as the expected trajectory of per-capita floor area. We will then develop a fully-fledged building energy and emissions model for modelling and analysing policy scenarios to investigate the trade-offs of embodied-versus-operational energy and carbon facing Chinese residential buildings.

Author Contributions: Conceptualization, W.Z. and A.M.; methodology, W.Z.; software, W.Z.; formal analysis, W.Z.; writing—original draft preparation, W.Z.; writing—review and editing, A.M., D.M.R., and P.G.; supervision, A.M., D.M.R., and P.G.

Funding: This research received no external funding.

Acknowledgments: W.Z. was funded by the Cambridge Trust.

Conflicts of Interest: The authors declare no conflicts of interest.

References

1. IEA. *World Energy Outlook 2015 Special Report on Energy and Climate Change*; OECD/IEA: Paris, France, 2015.
2. UNDP. Issue Brief: Fuel Cell Vehicle Development in China. 2015. Available online: http://www.cn.undp.org/content/china/en/home/library/environment_energy/issue-brief--fuel-cell-vehicle-development-in-china.html (accessed on 17 July 2018).
3. US EPA. Global Greenhouse Gas Emissions Data. 2017. Available online: https://www.epa.gov/ghgemissions/global-greenhouse-gas-emissions-data#Country (accessed on 20 July 2018).
4. World Bank. GEF to Support Implementation of China's Energy Efficiency and Environment Programs. 2017. Available online: http://www.worldbank.org/en/news/press-release/2017/03/16/gef-to-support-implementation-of-chinas-energy-efficiency-and-environment-programs (accessed on 20 July 2018).
5. IEA. World Energy Outlook 2013. 2013. Available online: http://www.oecd-ilibrary.org/energy/world-energy-outlook-2013_weo-2013-en (accessed on 20 July 2018).
6. IEA. World Energy Outlook 2014. 2014. Available online: http://www.oecd-ilibrary.org/energy/world-energy-outlook-2014_weo-2014-en (accessed on 20 July 2018).
7. IEA. World Energy Outlook 2015. 2015. Available online: http://www.oecd-ilibrary.org/energy/world-energy-outlook-2015_weo-2015-en (accessed on 20 July 2018).
8. IEA. *World Energy Outlook 2016*; OECD Publishing: Paris, France, 2016.
9. IEA. Building Energy Use in China: Transforming Construction and Influencing Consumption to 2050. 2015. Available online: https://www.iea.org/publications/freepublications/publication/PARTNERCOUNTRYSERIESBuildingEnergy_WEB_FINAL.pdf (accessed on 12 September 2018).
10. National Bureau of Statistics. *China Statistical Yearbook 2017*; China Statistics Press: Beijing, China, 2017. Available online: http://www.stats.gov.cn/tjsj/ndsj/2017/indexeh.htm (accessed on 19 January 2019).
11. Global Alliance for Buildings and Construction. Global Status Report 2016. 2016. Available online: http://www.globalabc.org/bundles/app/pdf/20161114_GABC-GSR-Report_Updated_Web-version.pdf (accessed on 13 July 2018).
12. IEA and IPEEC. Building Energy Performance Metrics: Supporting Energy Efficiency Progress in Major Economies. 2015. Available online: https://www.iea.org/publications/freepublications/publication/BuildingEnergyPerformanceMetrics.pdf (accessed on 25 July 2018).

13. IEA. Tracking Clean Energy Progress 2016. 2016. Available online: http://www.iea.org/publications/freepublications/publication/TrackingCleanEnergyProgress2016.pdf (accessed on 22 October 2018).

14. THUBERC. *2016 Annual Report on China Building Energy Efficiency*; China Architecture & Building Press: Beijing, China, 2016.

15. NDRC. Enhanced Actions on Climate Change: China's Intended Nationally Determined Contributions. 2015. Available online: http://www4.unfccc.int/ndcregistry/PublishedDocuments/ChinaFirst/China%27sFirstNDCSubmission.pdf (accessed on 25 July 2017).

16. Guan, D.; Meng, J.; Reiner, D.M.; Zhang, N.; Shan, Y.; Mi, Z.; Shao, S.; Liu, Z.; Zhang, Q.; Davis, S.J. Structural decline in China's CO2 emissions through transitions in industry and energy systems. *Nat. Geosci.* **2018**, *11*, 551–555. [CrossRef]

17. Huang, W. Speech at the Meeting of "She Hui Zhu Yi Xin Nong Cun Jian She", "Building New Countryside of the Socialist Society". 2006. Available online: http://www.mohurd.gov.cn/jsbfld/200612/t20061225_165486.html (accessed on 16 March 2019).

18. Hu, M.; Bergsdal, H.; van der Voet, E.; Huppes, G.; Muller, D.B. Dynamics of urban and rural housing stocks in China. *Build. Res. Inf.* **2010**, *38*, 301–317. [CrossRef]

19. China Daily. Short-Lived Buildings Create Huge Waste. 2010. Available online: http://www.chinadaily.com.cn/china/2010-04/06/content_9687545.htm (accessed on 25 May 2018).

20. Cai, W.; Wan, L.; Jiang, Y.; Wang, C.; Lin, L. Short-Lived Buildings in China: Impacts on Water, Energy, and Carbon Emissions. *Environ. Sci. Technol.* **2015**, *49*, 13921–13928. [CrossRef] [PubMed]

21. U.S. Department of Energy. 2011 Buildings Energy Data Book. 2012. Available online: https://catalog.data.gov/dataset/buildings-energy-data-book (accessed on 16 June 2018).

22. Hermann, A.; Deason, J.; Hobbs, A.; Novikova, A.; Yang, X.; Shengyuan, Z. Buildings Energy Efficiency in China, Germany, and the United States. 2013. Available online: http://climatepolicyinitiative.org/wp-content/uploads/2013/04/Buildings-Energy-Efficiency-in-China-Germany-and-the-United-States.pdf (accessed on 10 November 2017).

23. Yang, W.; Kohler, N. Simulation of the evolution of the Chinese building and infrastructure stock. *Build. Res. Inf.* **2008**, *36*, 1–19. [CrossRef]

24. Fawley, B.W.; Wen, Y. The Great Chinese Housing Boom, Econ. Synopses. 2013. Available online: http://research.stlouisfed.org/publications/es/article/9774 (accessed on 24 May 2017).

25. IEA. *Transition to Sustainable Buildings—Strategies and Opportunities to 2050*; OECD Publishing: Paris, France, 2013. [CrossRef]

26. Chalmers, P. Climate Change: Implications for Buildings. Key Findings from the Intergovernmental Panel on Climate Change Fifth Assessment Report. 2014. Available online: http://www.cisl.cam.ac.uk/business-action/low-carbon-transformation/ipcc-climate-science-business-briefings/pdfs/briefings/IPCC_AR5__Implications_for_Buildings__Briefing__WEB_EN.pdf (accessed on 16 August 2017).

27. Thomas, S. Energy Efficiency Policies for Buildings. 2015. Available online: http://www.bigee.net/media/filer_public/2015/02/06/bigee_broschuere_energy_efficiency_policy_in_buildings.pdf (accessed on 16 August 2017).

28. IPCC. Buildings. In *Climate Change 2014: Mitigation of Climate Change. Contribution of Working Group III to the Fifth Assessment Report of the Intergovernmental Panel on Climate Change*; 2014; Available online: https://www.ipcc.ch/pdf/assessment-report/ar5/wg3/ipcc_wg3_ar5_chapter9.pdf (accessed on 13 July 2017).

29. Institution of Civil Engineers. Embodied Energy and Carbon. 2015. Available online: https://www.ice.org.uk/knowledge-and-resources/briefing-sheet/embodied-energy-and-carbon (accessed on 20 July 2018).

30. Lin, L.; Jiang, Y.; Yan, D.; Chen, P. Analysis on building construction energy consumption and CO_2 emission in China. *China Energy* **2015**, *37*, 5–10. (In Chinese)

31. Pomponi, F.; Moncaster, A. Scrutinising embodied carbon in buildings: The next performance gap made manifest. *Renew. Sustain. Energy Rev.* **2018**, *81*, 2431–2442. [CrossRef]

32. Urge-Vorsatz, D.; Petrichenko, K.; Antal, M.; Staniec, M.; Labelle, M.; Ozden, E.; Labzina, E. Best Practice Policies for Low Energy and Carbon Buildings: A Scenario Analysis. 2012. Available online: http://www.gbpn.org/sites/default/files/08.CEUTechnicalReportcopy_0.pdf (accessed on 27 July 2017).

33. Hu, M.; Pauliuk, S.; Wang, T.; Huppes, G.; van der Voet, E.; Müller, D.B. Iron and steel in Chinese residential buildings: A dynamic analysis. *Resour. Conserv. Recycl.* **2010**, *54*, 591–600. [CrossRef]

34. Hu, M.; van der Voet, E.; Huppes, G. Dynamic Material Flow Analysis for Strategic Construction and Demolition Waste Management in Beijing. *J. Ind. Ecol.* **2010**, *14*, 440–456. [CrossRef]

35. Huang, T.; Shi, F.; Tanikawa, H.; Fei, J.; Han, J. Materials demand and environmental impact of buildings construction and demolition in China based on dynamic material flow analysis. *Resour. Conserv. Recycl.* **2013**, *72*, 91–101. [CrossRef]

36. Hong, L.; Zhou, N.; Feng, W.; Khanna, N.; Fridley, D.; Zhao, Y.; Sandholt, K. Building stock dynamics and its impacts on materials and energy demand in China. *Energy Policy* **2016**, *94*, 47–55. [CrossRef]

37. Shi, J.; Chen, W.; Yin, X. Modelling building's decarbonization with application of China TIMES model. *Appl. Energy* **2016**, *162*, 1303–1312. [CrossRef]

38. Zhou, N.; Fridley, D.; McNeil, M.; Zheng, N.; Ke, J.; Levine, M. China's Energy and Carbon Emissions Outlook to 2050. 2011. Available online: https://china.lbl.gov/sites/all/files/lbl-4472e-energy-2050april-2011.pdf (accessed on 7 August 2018).

39. Fridley, D.; Zheng, N.; Zhou, N.; Ke, J.; Hasanbeigi, A.; Price, L. *China Energy and Emissions Paths to 2030*, 2nd ed.; 2012. Available online: http://eta-publications.lbl.gov/sites/default/files/lbl-4866e-rite-modelaugust2012.pdf (accessed on 11 November 2018).

40. Zhou, N.; Fridley, D.; Khanna, N.Z.; Ke, J.; McNeil, M.; Levine, M. China's energy and emissions outlook to 2050: Perspectives from bottom-up energy end-use model. *Energy Policy* **2013**, *53*, 51–62. [CrossRef]

41. McNeil, M.A.; Feng, W.; Can, S.d.d.; Khanna, N.Z.; Ke, J.; Zhou, N. Energy efficiency outlook in China's urban buildings sector through 2030. *Energy Policy* **2016**, *97*, 532–539. [CrossRef]

42. Yu, S.; Eom, J.; Evans, M.; Clarke, L. A long-term, integrated impact assessment of alternative building energy code scenarios in China. *Energy Policy* **2014**, *67*, 626–639. [CrossRef]

43. Eom, J.; Clarke, L.; Kim, S.H.; Kyle, P.; Patel, P. China's building energy demand: Long-term implications from a detailed assessment. *Energy* **2012**, *46*, 405–419. [CrossRef]

44. Delmastro, C.; Lavagno, E.; Mutani, G. Chinese residential energy demand: Scenarios to 2030 and policies implication. *Energy Build.* **2015**, *89*, 49–60. [CrossRef]

45. Yang, T.; Pan, Y.; Yang, Y.; Lin, M.; Qin, B.; Xu, P.; Huang, Z. CO2 emissions in China's building sector through 2050: A scenario analysis based on a bottom-up model. *Energy* **2017**, *128*, 208–223. [CrossRef]

46. Shen, X. Has the per Capita Urban Residential Floor Area in China Exceeded 32 m2? The Biggest Lie of Chinese Livelihood, Tencent Housing. 2013. Available online: https://house.qq.com/bside/31.htm (accessed on 22 May 2019).

47. Liang, W. Is the Per Capita Urban Residential Floor Area in China 33 m^2 or 18 m^2? 2014. Available online: http://www.cssn.cn/preview/zt/19364/19369/201403/t20140313_1027648.shtml (accessed on 22 May 2019).

48. Ren, Z.; Xiong, C.; Bai, X. Urban Housing in China: Surplus or shortage? 2019. Available online: http://m.zqrb.cn/house/hangyedongtai/2019-02-18/A1550447505753.html (accessed on 22 May 2019).

49. Li, X.; Xu, D. Clarification of misled statistic data: Overestimated per-capita housing area. *China Econ. J.* **2013**, *6*, 134–151. [CrossRef]

50. THUBERC. *2017 Annual Report on China Building Energy Efficiency*; China Architecture & Building Press: Beijing, China, 2017.

51. Sterman, J.D. *Business Dynamics: Systems Thinking and Modelling for a Complex World*; Irwin McGraw-Hill: Boston, MA, USA, 2000.

52. Richardson, G.P. System Dynamics. In *Encyclopedia of Operations Research and Management Science*; Springer Science & Business Media B.V.: New York, NY, USA, 2001; pp. 807–810.

53. Shepherd, S.; Emberger, G. Introduction to the special issue: System dynamics and transportation. *Syst. Dyn. Rev.* **2010**, *26*, 193–194. [CrossRef]

54. Ventana Systems Inc. Vensim Software. 2019. Available online: https://vensim.com/vensim-software/ (accessed on 19 January 2019).

55. R Core Team. *R: A Language and Environment for Statistical Computing*; R Core Team: Vienna, Austria, 2019.

56. Allison, P. *Survival Analysis Using SAS: A Practical Guide*; SAS Institute: Cary, NC, USA, 2010.

57. Liu, X. *Survival Analysis: Models and Applications*; Wiley: Hoboken, NJ, USA, 2012. [CrossRef]

58. Miatto, A.; Schandl, H.; Tanikawa, H. How important are realistic building lifespan assumptions for material stock and demolition waste accounts? *Resour. Conserv. Recycl.* **2017**, *122*, 143–154. [CrossRef]

59. OECD. Measuring Capital: OECD Manual: Measurement of Capital Stocks, Consumption of Fixed Capital and Capital Services. 2001. Available online: https://www.oecd.org/std/na/1876369.pdf (accessed on 11 May 2017).

60. OECD. Measuring Capital: OECD Manual 2009. 2009. Available online: https://www.oecd.org/std/productivity-stats/43734711.pdf (accessed on 11 May 2017).

61. Johnstone, I.M. Energy and Mass Flows of Housing: Estimating Mortality. *Build. Environ.* **2001**, *36*, 43–51. [CrossRef]

62. Bohne, R.A.; Brattebø, H.; Bergsdal, H.; Hovde, P.J. Estimation of the Service Life of Residential Buildings, and Building Components, in Norway. In Proceedings of the City Surface of Tomorrow Conference, Vienna, Austria, 8–9 June 2006; pp. 29–33. Available online: https://www.academia.edu/27632589/Estimation_of_the_Service_Life_of_Residential_Buildings_and_Building_Components_in_Norway (accessed on 12 October 2018).

63. Müller, D.B. Stock dynamics for forecasting material flows—Case study for housing in The Netherlands. *Ecol. Econ.* **2006**, *59*, 142–156. [CrossRef]

64. Aksözen, M.; Hassler, U.; Rivallain, M.; Kohler, N. Mortality analysis of an urban building stock. *Build. Res. Inf.* **2016**, *45*, 1–19. [CrossRef]

65. Forbes, C.; Evans, M.; Hastings, N.; Peacock, B. *Statistical Distributions*, 4th ed.; John Wiley & Sons, Inc.: Hoboken, NJ, USA, 2011.

66. McLaren, C.; Stapenhurst, C. ONS Methodology Working Paper Series No 3: A Note on Distributions Used When Calculating Estimates of Consumption of Fixed Capital. 2015. Available online: http://webarchive.nationalarchives.gov.uk/20160111030849/http://www.ons.gov.uk/ons/guide-method/method-quality/specific/gss-methodology-series/ons-working-paper-series/index.html (accessed on 31 March 2018).

67. Rinne, H. *The Weibull Distribution: A Handbook*; Chapman and Hall/CRC: Boca Raton, FL, USA, 2008. [CrossRef]

68. Palisade. Risk Analysis and Simulation Add-In for Microsoft® Excel. 2016. Available online: https://www.palisade.com/downloads/documentation/75/EN/RISK7_EN.pdf (accessed on 31 March 2018).

69. Zaiontz, C. Weibull Distribution. 2018. Available online: http://www.real-statistics.com/other-key-distributions/weibull-distribution/ (accessed on 31 March 2018).

70. Liu, H.; Yang, F.; Xu, Y. Analysis on urban housing in China based on 2010 population census. *J. Tsinghua Univ. Philos. Soc. Sci.* **2013**, *6*, 138–147. (In Chinese)

71. THUBERC. *2012 Annual Report on China Building Energy Efficiency*; China Architecture & Building Press: Beijing, China, 2012.

72. THUBERC. *2019 Annual Report on China Building Energy Efficiency*; China Architecture & Building Press: Beijing, China, 2019.

Communication

Measuring Multi-Scale Urban Forest Carbon Flux Dynamics Using an Integrated Eddy Covariance Technique

Kaidi Zhang [1,2,†], Yuan Gong [3,†], Francisco J. Escobedo [4], Rosvel Bracho [5], Xinzhong Zhang [1,2] and Min Zhao [2,*]

1 School of Environmental and Geographical Sciences, Shanghai Normal University, Shanghai 200234, China
2 Research Center of Urban Ecology and Environment, Shanghai Normal University, Shanghai 200234, China
3 College of Biology and the Environment, Nanjing Forestry University, Nanjing 210037, China
4 Functional and Ecosystem Ecology Unit (EFE), Biology Program, Faculty of Natural Sciences and Mathematics, Universidad del Rosario, Bogotá D.C. 111221492, Colombia
5 School of Forest Resources and Conservation, University of Florida, Gainesville, FL 32611, USA
* Correspondence: zhaomin@shnu.edu.cn
† These authors contributed equally to this work.

Received: 14 July 2019; Accepted: 4 August 2019; Published: 11 August 2019

Abstract: The multi-scale carbon-carbon dioxide (C-CO_2) dynamics of subtropical urban forests and other green and grey infrastructure types were explored in an urbanized campus near Shanghai, China. We integrated eddy covariance (EC) C-CO_2 flux measurements and the Agroscope Reckenholz-Tänikon footprint tool to analyze C-CO_2 dynamics at the landscape-scale as well as in local-scale urban forest patches during one year. The approach measured the C-CO_2 flux from different contributing areas depending on wind directions and atmospheric stability. Although the study landscape was a net carbon source (2.98 Mg C ha^{-1} yr^{-1}), we found the mean CO_2 flux in urban forest patches was -1.32 μmol m^{-2}s^{-1}, indicating that these patches function as a carbon sink with an annual carbon balance of -5.00 Mg C ha^{-1}. These results indicate that urban forest patches and vegetation (i.e., green infrastructure) composition can be designed to maximize the sequestration of CO_2. This novel integrated modeling approach can be used to facilitate the study of the multi-scale effects of urban forests and green infrastructure on CO_2 and to establish low-carbon emitting planning and planting designs in the subtropics.

Keywords: carbon dioxide offsets; ART footprint tool; urban ecosystems; nature-based solutions; green infrastructure

---

1. Introduction

Urban areas are a complex matrix of different land covers, primarily including green infrastructure (GI; e.g., urban forests and other vegetation patches) and grey infrastructure (e.g., buildings, roads, and impervious). This matrix affects greenhouse gas (GHG) dynamics due to the combination of fossil fuel emissions from transportation and industrial activities and carbon-carbon dioxide (C-CO_2) sequestration by urban forests and other areas of natural and semi-natural vegetation, wetlands and rivers [1]. As such, there is increasing research on urban C-CO_2 dynamics and the use of urban forests as a nature-based solution across many cities of the world [2–4].

However, multi-scale modeling of urban C-CO_2 dynamics for different urban land covers, including urban and peri-urban forests in developing countries, is complex due to: Landscape heterogeneity, socioeconomic factors, emission sources, seasonality, wind direction-speed, and atmospheric stability [4,5]. Thus, many studies on the role of vegetation on C have used indirect methods such as allometric biomass equations and tree inventory data to assess C stocks and CO_2 sequestration in urban areas [6,7]. Other empirical approaches, such as urban metabolism [8] and direct measurements using eddy covariance (EC) flux observation systems [9], are also being used to assess these urban C-CO_2 dynamics. However, of these methods, EC allows for integrated measurements of C-CO_2 fluxes between the surface and the atmosphere within a contribution area to measured C fluxes, known as the footprint, which consists of sources and sinks [9–11]. These contribution areas are generally landscape-scale segments or portions within the study footprint [12]. As such, the C-CO_2 dynamics of smaller localized patches and areas within the footprint are much more difficult to model [4,10].

But, the available ART Footprint Tool (Agroscope Reckenholz-Tänikon; ART hereafter), based on the Kormann and Meixner model, is a relatively accessible approach for modeling source-sink heterogeneity and the partial flux contribution of specific local-scale areas inside a footprint [13]. Although little used in cities, it can potentially quantify how local-scale urban forest or other GI patches offset CO_2; a topic that is increasingly important to researchers and decision-makers globally. Such research using direct measurements of C-CO_2 flux source-sink characteristics and their use as a nature-based solution in subtropical urban forests in low-middle income countries is scarce [2,3,9].

Therefore, the aim of this short communication is to explore a novel approach for estimating the landscape and local-scale C-CO_2 flux dynamics of subtropical urban forests in an urbanizing area near subtropical Shanghai, China. Our specific objectives were two-fold. First, we quantified the overall CO_2 flux variation in an urbanized landscape and its different land covers. Second, we then explored the local-scale C-CO_2 flux dynamics of the specific urban forest, mixed, and grey infrastructure patches. We did so by combining Geographical Information Systems, direct annual EC measurements of C-CO_2 fluxes with the ART contribution area model. We then compared the C dynamics of urban forests with other areas of green-grey infrastructure [13].

As we will point out in the methods, the below approach requires advanced expertise, state of the art equipment, and data to implement. In particular, the long-term datasets required include canopy height and fine-scale meteorological and environment measurements, as well as spatial land use cover boundaries. Similarly, these local-scale flux source area assessments need CO_2 flux partitioning in different time series for the different land-use types. As such all these requirements are needed to make the below flux calculations and spatial assessments. That said, the novel, integrated, multi-scale techniques explored here could provide an improved approach to better understand, plan, manage, and design urban forests as essential components of urban nature-based solutions.

2. Materials and Methods

The 456-hectare study area, or landscape, is located at the Fengxian University Campus, approximately 40 km south of central Shanghai, China, in a coastal plain with a subtropical monsoon climate. Southeast winds dominate during the summer and northwest winds in the winter. The annual average temperature is 16 °C, the mean minimum and maximum temperatures are 4 °C in January and 33 °C in July–August, respectively [14,15]. The mean annual precipitation is 1200 mm and occurs mostly in the summer. Seasonality, as used in this study, is defined as follows, where Spring is the 1st of March–31st May, Summer is the 1st of June–31st August, Autumn is the 1st of September–30th November, while Winter is the 1st of December–28th February.

The grey infrastructure land use/covers (LULCs; e.g., building, roads, impervious) accounted for 39.25% of the study landscape. The buildings alone accounted for 24.31% of the total area and had an average height of 10 m. The GI or vegetation-dominated LULCs accounted for 53.2% of the study area, and the remaining was mixed and water bodies. Vegetation is characterized by a subtropical evergreen broad-leaved forest tree types dominated by *Cinnamomum camphora* with a canopy height of 8 m and other common herbaceous species such as *Acorus calamus* L., *Phragmites australis*, and *Ophiopogon japonicus*.

The EC system used for this short communication consisted of a three-dimensional ultrasonic anemometer (Windmaster, Gill, UK), and infrared gas analyzer (Li-7500, Licor, Lincoln, NE, USA), mounted on 20 m tall antenna tower in the center of the study area. High-frequency data was collected at 10 Hz on a CR3000 datalogger (Campbell Scientific Instruments, Logan, UT, USA) from April 2015 to March 2016. Carbon dioxide fluxes were calculated in 30 min intervals using EddyPro 5.1.1 software (Li-COR, Lincoln, NE, USA), which applied a coordinate rotation to adjust for local wind streamline and corrects for: The frequency response losses, spectral attenuation, and density fluctuations due to changes in air temperature and air moisture content.

The estimated fluxes were filtered following quality control indicators generated by the processing software (0-1-2; where "0" indicates best to "2", the poorest quality data) and the steady-state and developed turbulent conditions tests [16]. The fluxes were also eliminated under raining conditions or when less than 90% of the high-frequency data was collected in the 30 min intervals. Sixty-six percent of data remained after this filtering protocol was applied [17]. Atmospheric stability was defined according to Obukhov's stability lengths (L), where L < 0 represent unstable and L > 0 represent stable atmospheric stability, respectively [13,18]. Temporal gaps in the fluxes, usually less than 2 consecutive hours, were filled using the diurnal variation method (MDV) by interpolating the average value of the data generated at the same time in a cycle of 7 days for daytime and 14 days for night van 't Hoff model time [18–20]. For longer periods, the Landsberg model was used for daytime and CO_2 fluxes were related to photosynthetically active radiation, while the van 't Hoff model, which relates fluxes to soil temperature, was used for nighttime gap-filling [19,20]. The MDV method was used for all data gap filling procedures when estimating CO_2 fluxes of both the footprint and different GI areas within the study area. Negative CO_2 flux values indicate CO_2 uptake.

To attribute the CO_2 fluxes to their sources within the study landscape, we first estimated the characteristics and range of flux sources in the entire contributing area, or footprint, located on the urbanized campus [21,22]. The footprint had a radius of 500 m relative to the EC tower. We used the Hsieh model, based on the Lagrangian stochastic model and dimensional analysis, as it is suitable for long-term time series and requires few parameters, including: The observing height (m), zero-plane displacement value (m) and aerodynamic roughness length (m) [22]. Although the Flux—Source Area Model (FSAM) is also used for these types of analyses, it is much more complex and is not suitable for long-term flux footprint calculations [23,24].

Second, we estimated the flux contribution of 3 specific local-scale urban forest (i.e., tree dominated GI), mixed, and grey infrastructure patches to the total measured fluxes in the footprint using ART. We identified and selected the 3 specific patches using the following approach. A 15 × 15 m grid was initially established inside the footprint to spatially segregate the different LULCs. Then, 200 of these cells were grouped into either green or grey infrastructure types based on their LULCs. Afterward, the selected grey and green infrastructure cells were grouped into 26 different polygons that best represented the green and grey infrastructure patches across the 4 footprint quadrants (Figure 1). All polygons were summed producing the contribution of grey and GI to total measured fluxes and total fluxes from each quadrant [12]. The footprint function was then combined with the LULC data and the footprint contribution was calculated for 3 selected polygons that best represented the 3 different patch types (i.e., urban forest, mixed, and grey infrastructure) [12,13].

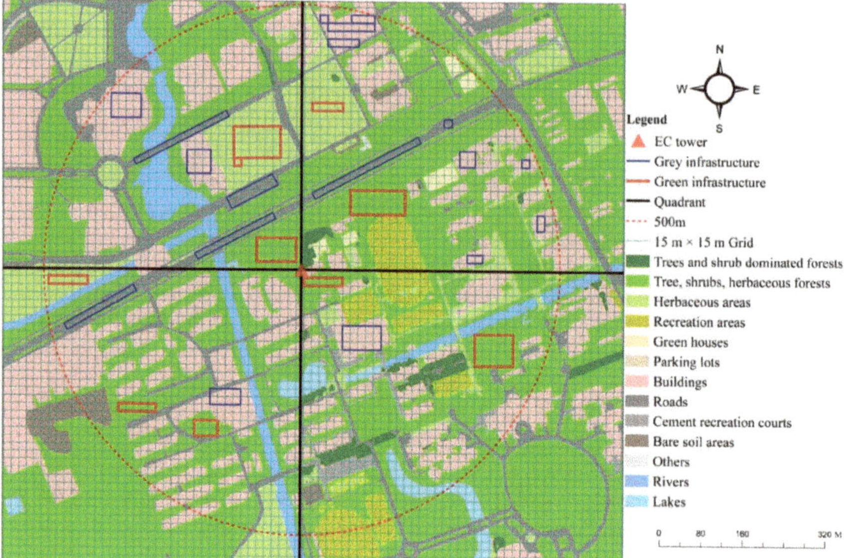

Figure 1. The eddy covariance footprint's grid, quadrants, and 26 polygons representing green (vegetation dominated cells) and grey infrastructure (building and impervious dominated cells) patches in the university campus study landscape near Shanghai, China.

3. Results

Ranges of prevailing seasonal wind direction and the percent of total frequency were estimated (Appendix A) and then the Hsieh footprint model was used to calculate the range of 90% of flux contribution area to the measured C-CO_2 flux according to the prevailing wind direction. The flux contribution area length for non-prevailing winds under stable atmospheric conditions was 260–810 m during the spring. When the atmosphere was unstable, the length of the flux contribution area was 121–220 m. The flux contribution areas for seasonal prevailing winds under stable and unstable atmospheric conditions are shown in Appendix A. As expected, all seasons exhibited a pattern of increased flux contribution areas as the atmospheric stability improved. The total daily C-CO_2 for the study is shown in Figure 2.

The total C-CO_2 balance in the study landscape was 2.98, with the northeast quadrant contributing 1.22, followed by the southwest with 0.78 and southeast and northwest quadrants contributing 0.47 and 0.51 Mg C ha^{-1} yr^{-1}, respectively. The sum of carbon emissions per quadrant accounts for 88% of the observed values in the total footprint.

The grey and GI areas were then selected throughout the footprint and we combined this EC data to characterize seasonal C-CO_2 fluxes (Figure 2) [12]. Overall, the annual carbon budget for GI areas was −5.00 Mg C ha^{-1} and the seasonal carbon budget in the spring, summer, autumn, and winter were: −1.52, −2.76, −1.02, and 0.30 Mg C ha^{-1} yr^{-1}, respectively. The annual average CO_2 flux for GI was −1.32 µmol m^{-2}s^{-1}. The annual carbon budget for the grey infrastructure areas was 8.17 Mg C/ha, and there were no obvious seasonal changes. The peak CO_2 flux values were within the range of 5–8 µmol m^{-2}s^{-1}. The CO_2 flux peak value decreased due to the lower traffic flow and reduced human activities at night. The annual average CO_2 flux for the grey infrastructure types was 2.16 µmol m^{-2}s^{-1}.

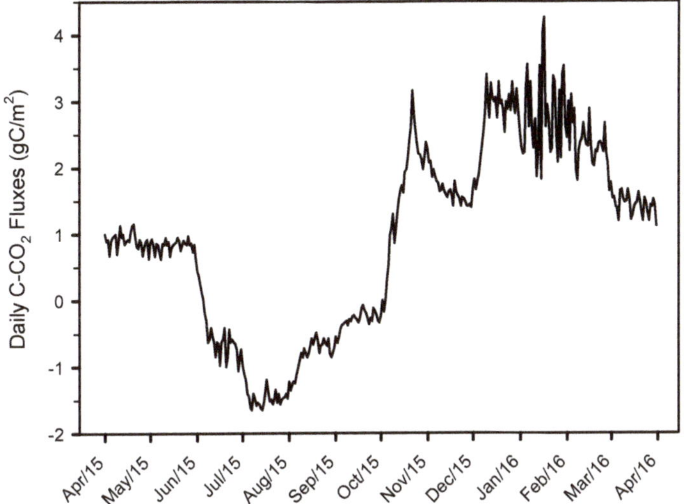

Figure 2. Daily C-CO$_2$ fluxes (gC m^{-2} day^{-1}) for the university campus study footprint near Shanghai, China from 1st April 2015 to 31st March 2016.

Parsing out the C-CO$_2$ flux of exclusively treed patches versus other areas of mixed vegetation is not possible as conventional EC studies the sum of all vegetation, soils, water, and other attributes in a given area [12]. Thus, hereafter we refer to patches of grey infrastructure and urban forests, or tree-dominated GI. Accordingly, we used three specific patches in the footprint of approximately 400 m^2 to represent: Urban forests, a mix of buildings and GI, and building-dominated grey infrastructure (Figure 3). An ideal annual average value was estimated according to the CO$_2$ flux in the three patches (Equation (1)) [18,23,25]. The F^i_c is the ideal value (Annual average, μmol m^{-2} s^{-1}), A_{Green} is the tree-dominated GI/urban forest area (ha), A_{Grey} is the area of the grey infrastructure (ha), and A_{test} is the study area (ha). The calculated annual average value is the calculated ART Footprint Tool CO$_2$ flux of the study area. The ideal and calculated values for urban forests, grey infrastructure, and mixed were −1.32 and −1.11, 2.16, and 2.48 and mixed were 0.42 and 0.95 μmol m^{-2}s^{-1}, respectively. By comparing the differences between the ideal and the calculated values, we determined the accuracy of the CO$_2$ fluxes for the different patches.

$$F^i_c = -1.32\frac{A_{Green}}{A_{test}} + 2.16\frac{A_{Grey}}{A_{test}} \tag{1}$$

The difference between the ideal and calculated value for the buildings plus GI patch was 0.53 μmol m^{-2}s^{-1} and differed greatly, relative to the difference between the ideal and calculated value for the vegetation and buildings. The difference of 0.53 μmol m^{-2}s^{-1} was within the acceptable error range, relative to the predicted CO$_2$ flux values [13].

Figure 3. Selected local-scale patches in the study landscape. From left to right are: Urban forests (100% trees and grass), grey infrastructure (100% building impervious), and a mixed patch (50% grey infrastructure, 50% green infrastructure).

4. Discussion and Conclusions

Our measured C-CO_2 fluxes were similar to other studies, despite the climatic and GI compositional differences [11]. Vesala et al. [12] found that vegetation in an area in Helsinki, Finland, had a C-CO_2 flux of less than 5 μmol m^{-2}s^{-1}. A CO_2 flux study in the nearby Chongming Island wetlands found that the distribution range under stable atmospheric conditions can reach 378 m [20,26]. The maximum winter CO_2 flux in Shenyang China can be 84 μmol m^{-2}s^{-1} [25], which, noticeably, is different from our grey infrastructure peak value and likely due to greater industrial and heating activities in cooler, northern, inland temperate China. Outside China, EC CO_2 fluxes for other cities are reported in Ward et al. [11].

In terms of the CO_2 flux contribution for different urban forest-GI types at the landscape scale, we found that the most notable difference was between herbaceous-woody and building-road LULCs; a finding similar to Yao et al., who identified forest cover in the Shanghai, China area as being a driver of C stocks [6]. The contribution of other LULC types in terms of CO_2 flux contribution was, however, lower as observed in other studies [5]. Thus, our results show the cumulative influence of these urban forests and other green infrastructure can have on CO_2 flux, as opposed to grey infrastructure, on the overall CO_2 flux contribution area [2,4,10].

We do note that the main limitation of this short communication is that we used one year of data and we did not analyze water and bare soil LULC types. These areas of "blue and brown" infrastructures are common in urban areas and will likely influence the C-CO_2 balance in the study area. But, our short communication presents an approach for future research, such as understanding how water bodies, wetlands, and urban development and construction activities affect the C flux of urban ecosystems and in improving the ART footprint tool functionality to better analyze urban ecosystems. Future studies can use the approach presented here and multiple years of data to better characterize the CO_2 budget for other GI types, such as areas planted with native versus non-native vegetation and other high versus low maintenance urban vegetation types. In particular, characterizing their seasonal and annual budgets across time and space could lead to better estimates of their true CO_2 offsetting potential.

Multi-scale studies combining geospatial, EC CO_2 flux data, and footprint/contribution area models are not common for subtropical, coastal, cities in low-middle income countries. Thus, this short communication demonstrates how long-term, local-scale data, and the ART tool can provide a more accessible and novel approach for modeling urban forest C-CO_2 dynamics and heterogeneity in cities. Such information can help us better urban plan, design, and justify the use of urban forests as integral components of nature-based solutions, which are becoming popular in many cities. The research approach presented in the paper can be used to quantify the role of low-carbon cities and local-scale effects of regulating greenhouse gas emissions. This study in subtropical China can also provide a baseline and approach for carbon flux research of the urban environment in other cities in the tropics and subtropics.

Author Contributions: Conceptualization, K.Z., Y.G. and M.Z.; methodology, K.Z. and Y.G.; software, K.Z. and Y.G.; validation, K.Z., Y.G. and M.Z.; formal analysis, Y.G.; investigation, K.Z. and Y.G.; resources, F.E.; data curation, K.Z., Y.G. and X.Z.; writing—original draft preparation, K.Z. and Y.G.; writing—review and editing, F.E., R.B. and M.Z.; visualization, Y.G.; supervision, M.Z.; project administration, M.Z.; funding acquisition, M.Z.

Funding: This research was funded by the National Natural Science Foundation of China, grant number 31100354 and The APC was funded by National Natural Science Foundation of China.

Acknowledgments: This research was funded by the National Natural Science Foundation of China (No. 31100354). We thank Jiemin Wang from the Northwest Institute of Eco-Environment and Resources, Chinese Academy of Sciences for the guidance on the use of footprint models.

Conflicts of Interest: The authors declare no conflict of interest.

Appendix A

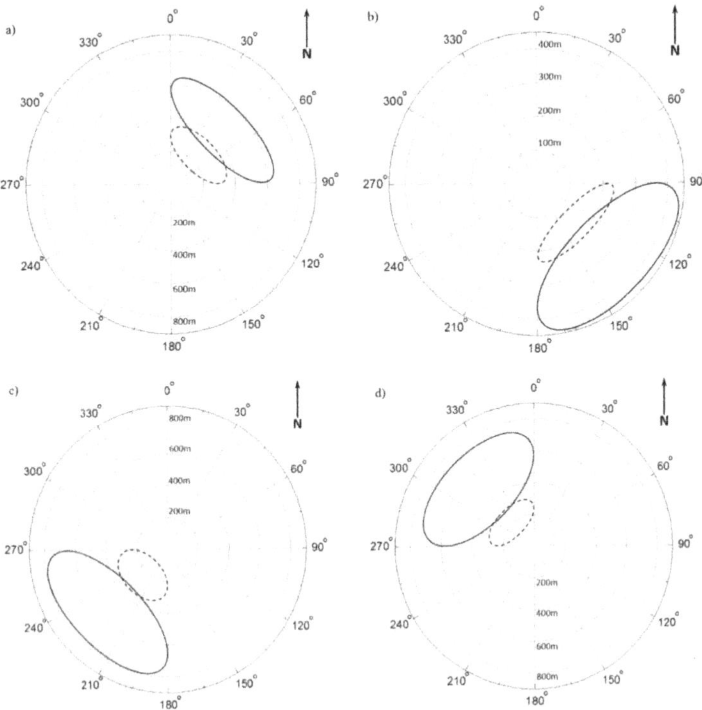

Figure A1. The farthest point distribution of 90% contribution area for dominate wind directions for different seasons in the Shanghai Fengxian University Campus. (**a**) Spring; (**b**) summer; (**c**) autumn; (**d**) winter. Solid lines represent stable, dotted line represents unstable atmospheric conditions.

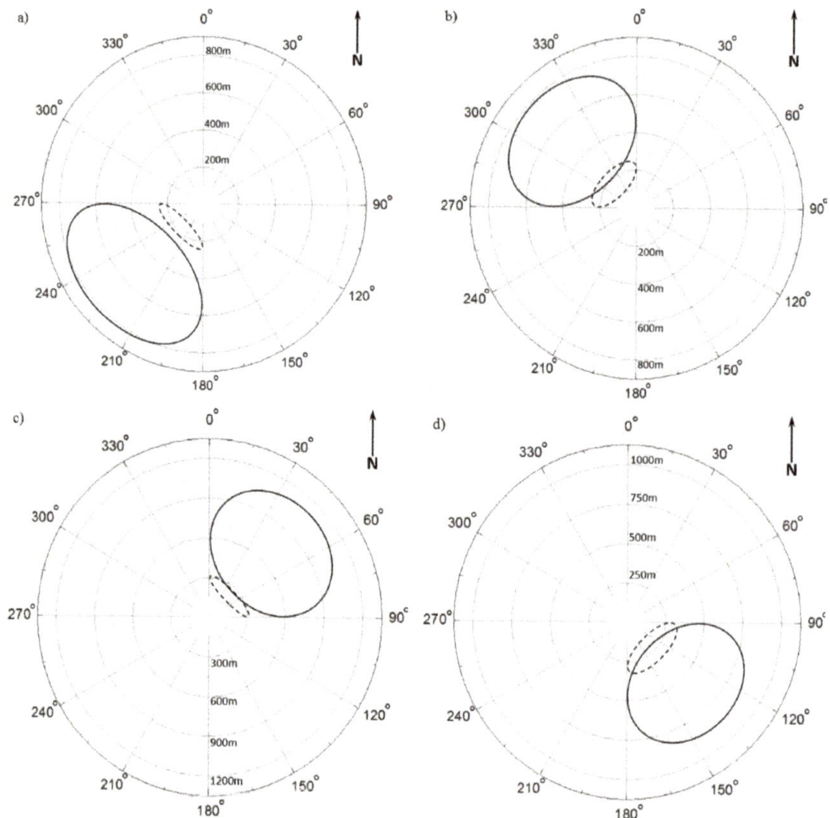

Figure A2. The Farthest point distribution of 90% contribution area for non-dominant wind directions during four different seasons in the Shanghai Fengxian University Campus. (**a**) Spring; (**b**) summer; (**c**) autumn; (**d**) winter. Circular shapes with solid lines represent stable, while dotted line represents unstable, atmospheric conditions.

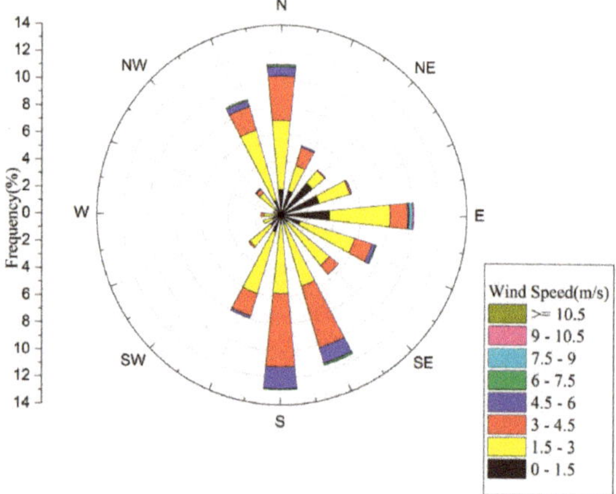

Figure A3. Wind rose for the Shanghai Fengxian University Campus study landscape from April 2015 to March 2016.

References

1. Xie, J.; Jia, X.; He, G.; Zhou, C.; Yu, H.; Wu, Y.; Bourque, C.P.A.; Liu, H.; Zha, T. Environmental control over seasonal variation in carbon fluxes of an urban temperate forest ecosystem. *Landsc. Urban Plan.* **2015**, *142*, 63–70. [CrossRef]
2. Briber, B.M.; Hutyra, L.R.; Dunn, A.L.; Raciti, S.M.; Munger, J.W. Variations in atmospheric CO_2 mixing ratios across a Boston, MA urban to rural gradient. *Land* **2013**, *2*, 304–327. [CrossRef]
3. Escobedo, F.J.; Giannico, V.; Jim, C.Y.; Sanesi, G.; Lafortezza, R. Urban Forests, Ecosystem Services, Green Infrastructure and Nature-Based Solutions: Nexus or Evolving Metaphors. *Urban For. Urban Green.* **2019**, *37*, 3–12. [CrossRef]
4. Guidolotti, G.; Calfapietra, C.; Pallozzi, E.; De Simoni, G.; Esposito, R.; Mattioni, M.; Nicolini, G.; Matteucci, G.; Brugnoli, E. Promoting the potential of flux-measuring stations in urban parks: An innovative case study in Naples, Italy. *Agric. For. Meteorol.* **2017**, *233*, 153–162. [CrossRef]
5. Gioli, B.; Gualtieri, G.; Busillo, C.; Calastrini, F.; Zaldei, A.; Toscano, P. Improving high resolution emission inventories with local proxies and urban eddy covariance flux measurements. *Atmos. Environ.* **2015**, *115*, 246–256. [CrossRef]
6. Yao, X.; Zhao, M.; Escobedo, F.J. What causal drivers influence carbon storage in Shanghai, China's Urban and Peri-Urban Forests. *Sustainability* **2017**, *9*, 577. [CrossRef]
7. Zhao, M.; Kong, Z.H.; Escobedo, F.J.; Gao, J. Impacts of urban forests on offsetting carbon emissions from industrial energy use in Hangzhou, China. *J. Environ. Manag.* **2010**, *91*, 807–813. [CrossRef] [PubMed]
8. Barles, S. Urban metabolism of Paris and its region. *J. Ind. Ecol.* **2009**, *13*, 898–913. [CrossRef]
9. Velasco, E.; Perrusquia, R.; Jiménez, E.; Hernández, F.; Camacho, P.; Rodríguez, S.; Retama, A.; Molina, L.T. Sources and sinks of carbon dioxide in a neighborhood of Mexico City. *Atmos. Environ.* **2014**, *97*, 226–238. [CrossRef]
10. Awal, M.A.; Ohta, T.; Matsumoto, K.; Toba, T.; Daikoku, K.; Hattori, S.; Hiyama, T.; Park, H. Comparing the carbon sequestration capacity of temperate deciduous forests between urban and rural landscapes in central Japan. *Urban For. Urban Green.* **2010**, *9*, 261–270. [CrossRef]
11. Ward, H.C.; Kotthaus, S.; Grimmond, C.S.B.; Bjorkegren, A.; Wilkinson, M.; Morrison, W.T.J.; Evans, J.G.; Morison, J.I.L.; Iamarino, M. Effects of urban density on carbon dioxide exchanges: Observations of dense urban, suburban and woodland areas of southern England. *Environ. Pollut.* **2015**, *198*, 186–200. [CrossRef] [PubMed]
12. Vesala, T.; Järvi, L.; Launiainen, S.; Sogachev, A.; Rannik, Ü.; Mammarella, I.; Siivola, E.; Keronen, P.; Rinne, J.; Riikonen, A.; et al. Surface–atmosphere interactions over complex urban terrain in Helsinki, Finland. *Tellus Ser. B-Chem. Phys. Meteorol.* **2008**, *60*, 188–199. [CrossRef]
13. Neftel, A.; Spirig, C.; Ammann, C. Application and test of a simple tool for operational footprint evaluations. *Environ. Pollut.* **2008**, *152*, 644–652. [CrossRef] [PubMed]
14. Gong, Y.; Zhao, M.; Yao, X.; Guo, Z.J.; He, Y.; Zhang, L.P. Analysis and comparison of carbon flux contribution zones in urban ecological system based on Hsieh and Kljun models. *J. Environ. Eng. Technol.* **2017**, *7*, 225–231. (In Chinese)
15. Gong, Y.; Guo, Z.J.; Zhang, K.D.; Xu, L.; Wei, Y.Y.; Zhao, M. Impact of vegetation on CO_2 flux of a subtropical urban ecosystem. *Acta Ecol. Sin.* **2019**, *39*, 530–541.
16. Mauder, M.; Foken, T. Documentation and instruction manual of the eddy covariance software package TK2. *Univ. Bayreuth Abt. Mikrometeorologie* **2004**, *26*, 1614–8916.
17. Finkelstein, P.L.; Sims, P.F. Sampling error in eddy correlation flux measurements. *J. Geophys. Res. Atmos.* **2001**, *106*, 3503–3509. [CrossRef]
18. Gong, Y. A Study on Spatial Differentiation of Carbon Flux Based on Footprint Model A Case Study of Fengxian University City in Shanghai. Master Dissertation, Shanghai Normal University, Shanghai, China, 2017. (In Chinese).
19. Falge, E.; Baldocchi, D.; Olson, R.; Anthoni, P.; Aubinet, M.; Bernhofer, C.; Granier, A. Gap filling strategies for defensible annual sums of net ecosystem exchange. *Agric. For. Meteorol.* **2001**, *107*, 43–69. [CrossRef]
20. Gu, Y.J.; Gao, Y.; Guo, H.Q.; Zhao, B. Footprint Analysis for carbon flux in the wetland ecosystem of Chongming Dongtan. *J. Fudan Univ.* **2008**, *47*, 374–379. (In Chinese)

21. Liu, Y.J.; Hu, F.; Cheng, X.L.; Song, Z.P. Distribution of the source area and footprint of Beijing. *Chin. J. Atmos. Sci.* **2014**, *38*, 1044–1054. (In Chinese)
22. Hsieh, C.I.; Katul, G.; Chi, T.W. An approximate analytical model for footprint estimation of scalar fluxes in thermally stratified atmospheric flows. *Adv. Water Resour.* **2000**, *23*, 765–772. [CrossRef]
23. Guo, Z.J. Carbon Footprint Flux Analysis of Urban Underlying Surface: A Case Study of Fengxian University City in Shanghai. Master Dissertation, Shanghai Normal University, Shanghai, China, 2018. (In Chinese).
24. Guo, Z.J.; Gong, Y.; Zhang, K.D.; Zhang, L.P.; He, Y.; Xu, L.; Zhao, M. CO_2 flux footprints of impervious layer on complex land surface: A case study at the Fengxian College Park, Shanghai. *Acta Sci. Circumstantiae* **2018**, *38*, 772–779. (In Chinese)
25. Jia, Q.Y.; Zhou, G.S.; Wang, Y. Footprint characteristics of CO_2 flux over the urban district of Shenyang. *Acta Sci. Circumstantiae* **2010**, *30*, 1682–1687. (In Chinese)
26. Wang, J.T.; Zhou, J.H.; Ou, Q.; Zhong, Q.C.; Wang, K.Y. CO_2 flux footprint analysis of coastal polder wetlands in Dongtan of Chongming. *J. Ecol. Rural Environ.* **2014**, *30*, 588–594. (In Chinese)

Article

Does the Low-Carbon Pilot Initiative Reduce Carbon Emissions? Evidence from the Application of the Synthetic Control Method in Guangdong Province

Xuan Yu [1,*], Manhong Shen [1,2,*], Di Wang [1] and Bernadette Tadala Imwa [3]

1   Business School, Ningbo University, Ningbo 315211, China
2   Donghai Institute, Ningbo University, Ningbo 315211, China
3   Law School, Ningbo University, Ningbo 315211, China
*   Correspondence: 1511091887@nbu.edu.cn (X.Y.); smh@nbu.edu.cn (M.S.); Tel.: +86-1889-262-5280 (X.Y.)

Received: 24 May 2019; Accepted: 18 July 2019; Published: 23 July 2019

Abstract: As the world's top energy consumer and carbon emitter, China's carbon emissions policies, including the low-carbon pilot initiative (LCPI) implemented in July 2010, have important effects on global climate change. Therefore, accurately assessing the effect of this policy has become extremely important for low-carbon development. This article analyses the impact of implementing LCPI on regional carbon emissions by using Guangdong Province as the study area, which has the largest economic scale, population size and carbon emissions amongst China's low-carbon pilot provinces. The results suggest that for the entire 2010–2015 period, Guangdong's carbon emissions were reduced by about 10% due to the implementation of LCPI. This policy produced a significant impact on the carbon emissions from manufacturing industries but showed minimal impact on the carbon emissions from energy production. Unlike previous researchers who relied on estimations, the authors of this work obtained unified carbon emissions data for 1997–2015 from the China Emission Accounts and Datasets and then constructed comparison groups by using the synthetic control method instead of performing a subjective selection. The authors also examined the impact of LCPI on carbon emissions from different sources. This article proposes that policy support and low-carbon action are necessary for reducing regional carbon emissions and that the policies must be constantly adjusted during their implementation. The successful experiences in low-carbon pilots are also worth exploring and promoting in other regions.

Keywords: low-carbon pilot initiative; carbon emissions; policy effect evaluation; synthetic control method

1. Introduction

Over the past century, the global average temperature increased by 0.4–0.8 °C, which is expected to increase further to 1.4–5.8 °C over the next 100 years. Global warming has created many environmental problems, such as the melting of bipolar glaciers, rising sea levels, increased frequency of extreme weather and ecological damage, thereby turning this phenomenon into one of the major challenges being faced by the world today [1–4]. Scientific studies have confirmed a direct relationship between greenhouse gas emissions (e.g., carbon dioxide) and global climate change [5–8]. In addition, the amount of global carbon dioxide emissions continues to increase despite the calls of climate scientists and international organizations, including the United Nations, for cuts in energy consumption [9]. The research of Global Carbon Project reveals that the amount of global CO_2 emissions has increased by 1.6% in 2016–2017 and is expected to increase by 2.7% in 2018 [10].

China is the world's top energy consumer and CO_2 emitter [11]. Since its accession to the WTO in 2001, China's GDP growth rate has reached an average of 10% a year, thereby turning China into the world's second largest economy. However, the rapid development of China's economy comes at the cost of consuming huge amounts of resources. According to the National Bureau of Statistics of China, the total energy consumption of China increased by nearly three times from 1555 million tce (ton of standard coal equivalent) in 2001 to 4490 million tce in 2017 [12]. Such large quantification energy consumption also results in huge amounts of carbon emissions. Previous studies reveal that 27% of global carbon emissions in 2017 originated from China [10].

China and other countries around the world have reached a consensus that promoting a low-carbon economy with the goal of reducing greenhouse gas emissions can effectively solve those problems driven by climate change [13–16]. In 2004, the National Development Reform Commission (NDRC) promulgated the *Medium and Long-Term Special Plan for Energy Conservation* (2004) to reduce energy intensity [17]. Subsequently, China's top administrative institution, the State Council, promulgated *China's National Climate Change Program* (2007) and *White Paper on China's Actions and Strategy on Climate Change* (2008) to reduce the country's greenhouse gas emissions [18,19]. In 2009, the State Council announced its target to reduce the country's carbon intensity by 40%–45% by 2020 compared to its 2005 level [20], and this target was subsequently included in the *National 12th Five Year Plan* in 2010 [21]. To achieve such target, in July 2010, the NDRC started implementing its low-carbon pilot initiative (LCPI) in the five provinces of Liaoning, Hubei, Guangdong, Yunnan, and Shaanxi.

The implementation of LCPI has motivated many researchers to examine the effect of this policy. For example, Chu (2017), Deng (2017), and Dai (2015) argued that the implementation of LCPI has significantly inhibited regional carbon emissions and that this effect continues to increase every year. [22–24]. However, Lu (2017) argued that not all pilot areas were influenced by LCPI, with only Shaanxi showing significant reductions in its carbon emissions [25]. Similarly, Zhao (2018) argued that the effect of LCPI varies across different provinces. For instance, this policy effectively reduced the agricultural carbon emissions in Shaanxi, Hubei, and Liaoning but did not produce any obvious effect in Guangdong and Yunnan [26].

Previous studies have evaluated the effectiveness of LCPI, but their conclusions greatly vary and even contradict one another because of the differences in their data and methodologies. For instance, the most important outcome variable used in these studies (CO_2 emissions) was estimated by using different methods. However, emissions estimates are relatively uncertain in many circumstances and show large gaps [27]. The methodologies employed in previous research, including the Differences-in-Differences method, also show some limitations in evaluating policy effects, with researchers often selecting their comparison groups subjectively [28]. In addition, previous researches do not study deeply the impact of the LCPI on carbon emissions from different sources.

In this study, the authors obtained annual CO_2 emissions data at the provincial level for 1997–2015 from China Emission Accounts and Datasets (CEADs), which was constructed in 2017 and is considered by far the most authoritative and unified carbon emission dataset for emission-related research in China. The authors also used Guangdong Province as the study area given that this province has the largest economic scale, population size and carbon emissions amongst China's low-carbon pilot provinces. Data-driven procedures were also employed to construct suitable comparison groups based on the synthetic control method proposed by Abadie to examine whether LCPI has actually reduced the regional carbon emissions of China. The impact of this policy on carbon emissions from different sources has also been investigated.

2. Materials and Methods

2.1. Study Area

Guangdong Province, one of China's first low-carbon pilot provinces, is located in the southeastern part of the country (109°39'~117°19' E and 20°13'~25°31' N) facing the South China Sea (Figure 1). The Guangdong provincial administrative area includes 21 cities with a total population of 117 million, thereby making this area the most populated province in China. Since 1989, Guangdong's GDP has ranked first in the country, thereby turning Guangdong into China's largest economic province. In 2017, the total GDP of Guangdong amounted to 89,705.23 billion yuan or 10.8% of China's total GDP. The high amount of economic aggregates and total population also hugely contribute to the carbon emissions of this province. In fact, Guangdong's annual CO_2 emissions are higher than those of the other four low-carbon pilot provinces since 2001 (Figure 2a).

Figure 1. Location of Guangdong Province and its cities.

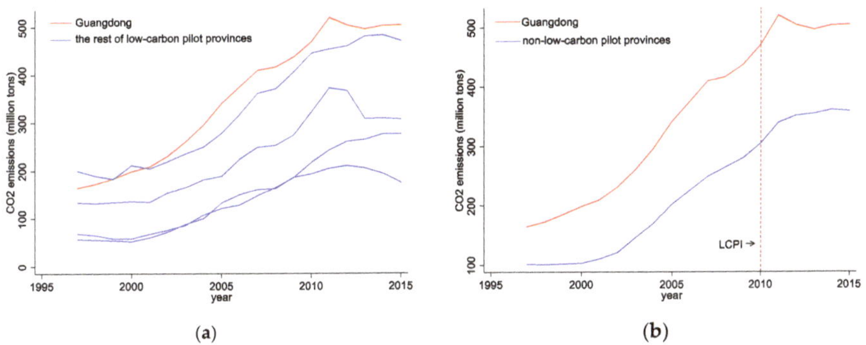

Figure 2. Comparison of annual CO_2 emissions between Guangdong and other provinces. (**a**) Comparison of annual CO_2 emissions between Guangdong and the rest of low-carbon pilot provinces, and (**b**) comparison of annual CO_2 emissions between Guangdong and non-low-carbon pilot provinces.

2.2. Policy Background

The main sources of carbon emissions in Guangdong are its energy production and manufacturing industries, which altogether account for 80.4% of the province's total emissions (Figure 3). The Guangdong government divided its implementation of LCPI into four periods, namely, the start-up period (August 2010–December 2010), the initial period (2011), the crucial period (2012–2013) and the deepening period (2014–2015), all of which aimed to reduce the CO_2 emissions of the province by 35% in 2015 from its 2005 level. To achieve this goal, Guangdong focuses on the optimization of its energy structure and the development of its low-carbon industry.

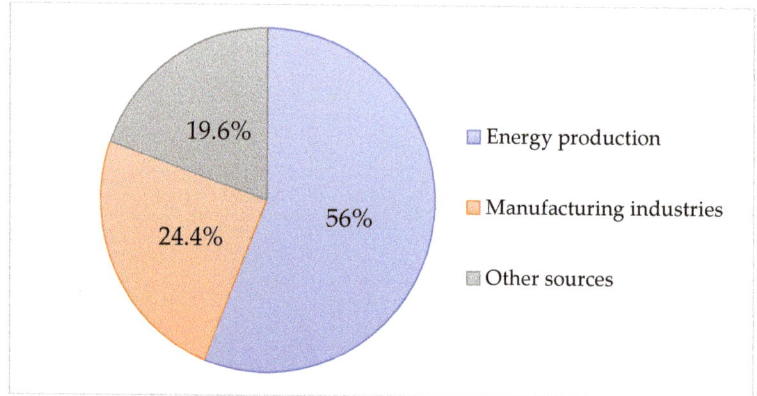

Figure 3. Source of carbon emissions in Guangdong Province.

To optimize an energy structure, the production of thermal power, nuclear power, and wind power must be promoted, the utilization of solar energy must be enhanced, biomass energy needs to be developed in moderation, a new type of rural energy must be developed according to the local conditions and other emerging energy sources, such as geothermal energy, tidal energy, and hydrogen energy, need to be cultivated. In addition, the government advocates that enterprises should pay attention to their energy conservation and improve their energy efficiency in their production processes. To create a low-carbon industry, the development of advanced manufacturing (e.g., nuclear power equipment, wind power equipment, automobile manufacturing, petrochemicals, shipbuilding and marine engineering equipment manufacturing) and strategic emerging manufacturing (e.g., semiconductor lighting, new energy vehicles, solar photovoltaic energy, energy conservation, and environmental protection) must be accelerated, whilst the development of modern service industries (e.g., finance, logistics, information services, technology services, outsourcing services, headquarters economies, business exhibitions, cultural creativity, and tourism) needs to be optimized. A circular economy must also be developed, clean production processes and technologies need to be promoted, waste generation from sources must be reduced and waste recycling must be encouraged.

In addition, Guangdong has also implemented some innovative policies. In institutional reform, the Guangdong Provincial Leading Group for Climate Change and Energy Conservation and Emission Reduction was established and directly led by the governor. Meanwhile, the low-carbon targets was incorporated into the performance evaluation system of government officials. In market mechanism, Guangdong pioneered the development of an emissions trading scheme (ETS), which could allow carbon emissions allowances to be traded. Unlike other areas in China, which use the free allocation of permits to reduce the financial burden on enterprises, Guangdong adopted a tougher approach by requiring compliance enterprises to buy allowances through auction. For example, the compliance enterprises in the iron and steel industry could receive a 97% quota for free only if they purchase 3% of allowances. The government also set a high reserve price of RMB 60/ton for mandatory auction. This allocation approach is superior to free allocation because it contributes to the identification of the proper price for carbon emissions, as well as raising money for the provincial low-carbon development fund [29].

Figure 2b plots the trends in the annual CO_2 emissions of Guangdong and the other provinces that are not covered by LCPI. As can be seen in this figure, before the implementation of LCPI, the trends in the annual CO_2 emissions of Guangdong and the other provinces were all showing a continuous growth. These trends began to diverge in 2011 when Guangdong's annual CO_2 emissions peaked and began to decline whereas those of the other provinces continued to increase. To evaluate the impact of LCPI on the carbon emissions of Guangdong, the following question needs to be addressed: If the LCPI was not implemented in 2010, how will Guangdong's carbon emissions evolve? This counterfactual can be estimated by using the synthetic control method.

2.3. Synthetic Control Method

The thinking of synthetic control method is to use Guangdong as the treatment group, find the appropriate weight through the predicted variables, and weight the average of the provinces where the LCPI has not been implemented. Then we synthesize a virtual Guangdong that has not implemented the LCPI and use it as a comparison group, and make the characteristics of the predicted variables in synthetic Guangdong and real Guangdong before the LCPI implementation as similar as possible. Finally, the difference of carbon emissions between real Guangdong and synthetic Guangdong after the LCPI implementation can be compared.

Suppose that we observe the carbon emissions of $J+1$ provinces for T years, and only the first province, that is Guangdong, implemented LCPI, hence the J remaining regions can be the control group. Suppose that T_0 is the year of the LCPI implementation, Y_{it}^N is the carbon emissions that could be observed for province i at time t in the absence of the LCPI, and Y_{it}^I is the carbon emissions that could be observed for province i at time t if province i is exposed to the implementation of the LCPI in periods T_{0+1} to T. Let Y_{it} be the actual carbon emissions that would be observed for province i at time t. So $\alpha_{it} = Y_{it}^I - Y_{it}^N$ is the effect of the implementation of LCPI for province i at time t, $\alpha_{1t} = Y_{1t}^I - Y_{1t}^N$ is the effect of LCPI on carbon emissions in Guangdong. Because all provinces' carbon emissions are not affected by LCPI before T_0, when $t \leq T_0$, $Y_{it}^I = Y_{it}^N$; when $T_0 < t \leq T$, $Y_{it}^I = Y_{it}^N + \alpha_{it}$. We introduce the dummy variable D_{it}. The observed actual carbon emissions for province i at time t is $Y_{it} = Y_{it}^N + D_{it}\alpha_{it}$, when $t \leq T_0$, $\alpha_{1t} = 0$; when $t > T_0$, $\alpha_{1t} = Y_{1t}^I - Y_{1t}^N = Y_{1t} - Y_{1t}^N$, Y_{1t} is the actual carbon emissions of Guangdong that can be observed after the implementation of the LCPI. We can estimate α_{1t} by estimating Y_{1t}^N. Suppose that:

$$Y_{it}^N = \delta_t + \theta_t Z_i + \lambda_t \mu_i + \varepsilon_{it} \tag{1}$$

where δ_t is an unknown common factor, Z_i are observed covariates which not affected by the LCPI, θ_t are unknown parameters, λ_t are unobserved common factors, μ_i are unknown factor loadings, and ε_{it} are error terms. Suppose that a weights vector W to estimate Y_{it}^N, $W = (w_2, w_3, \cdots, w_{J+1})$, $w_j \geq 0$, $w_2 + \cdots + w_{J+1} = 1$. The carbon emissions for each synthetic control indexed by W is

$$\sum_{j=2}^{J+1} w_j Y_{jt}^N = \delta_t + \theta_t \sum_{j=2}^{J+1} w_j Z_J + \lambda_t \sum_{j=2}^{J+1} w_j \mu_j + \sum_{j=2}^{J+1} w_j \varepsilon_{jt} \tag{2}$$

Suppose that there are $\left(w_2^*, \cdots, w_{J+1}^*\right)$ such that

$$\sum_{j=2}^{J+1} w_j^* Y_{j1} = Y_{11}, \sum_{j=2}^{J+1} w_j^* Y_{j2} = Y_{12}, \cdots, \sum_{j=2}^{J+1} w_j^* Y_{jT_0} = Y_{1T_0} \text{ and } \sum_{j=2}^{J+1} w_j^* Z_j = Z_1 \tag{3}$$

If $\sum_t^{T_0} \lambda_t' \lambda_t$ is nonsingular, then

$$Y_{it}^N - \sum_{j=2}^{J+1} w_j^* Y_{jt} = \sum_{j=2}^{J+1} w_j^* \sum_{s=1}^{T_0} \lambda_t \left(\sum_{n=1}^{T_0} \lambda_n' \lambda_n\right)^{-1} \lambda_s' \left(\varepsilon_{js} - \varepsilon_{1s}\right) - \sum_{j=2}^{J+1} w_j^* \left(\varepsilon_{js} - \varepsilon_{1s}\right) \tag{4}$$

Abadie proved that, the mean of the right-hand side of Equation (4) will be close to zero under standard conditions, hence we can use $\hat{\alpha}_{1t} = Y_{1t} - \sum_{j=2}^{J+1} w_j^* Y_{jt}$ for $t \in \{T_0 + 1, \cdots, T\}$ as an estimator of α_{1t}.

2.4. Variables and Data

The annual CO_2 emissions at the provincial level were used as the outcome variable and were collected from CEADs. Following Wang (2010), several variables were used as predictors of carbon emissions, including economic size, economic structure, economic level, population size, energy intensity, length of transport routes, and number of vehicles [30]. The details of these variables are shown in Table 1.

Table 1. Predictors of carbon emissions.

Predictor	Index	Abbreviation	Data Source
Economic size	GDP	GDP	China Statistical Yearbook
Economic structure	Secondary industry output/GDP	ES	China Statistical Yearbook
Economic level	Per capita GDP	PGDP	China Statistical Yearbook
Population size	Total population	POP	China Statistical Yearbook
Energy intensity	Energy consumption	EC	China Energy Statistics Yearbook
Length of transport routes	Total mileage of roads and railways	TD	China Transportation Yearbook
Number of vehicles	Number of civil cars	VN	China Transportation Yearbook

The synthetic control method was used to create a synthetic version of Guangdong that mirrors the values of the predictors of the real Guangdong before the LCPI implementation. Not all the other provinces (excluding Guangdong) were placed in the control group. Firstly, Tibet, Xinjiang, and Taiwan were eliminated from the sample because of missing data. Secondly, Liaoning, Hubei, Yunnan, and Shaanxi were eliminated as they also implemented LCPI in 2010. Thirdly, Hainan was eliminated given that this province was named as a low-carbon pilot province in 2012. Fourthly, Zhejiang was eliminated because LCPI includes not only provinces but also cities. The annual CO_2 emissions of Hangzhou, Ningbo, and Wenzhou, which are included amongst China's low-carbon pilot cities, are more than half of that of Zhejiang Province. Therefore, although not a low-carbon pilot province, Zhejiang is also greatly affected by LCPI.

After creating a synthetic Guangdong, the effect of LCPI on carbon emissions was estimated as the difference in the annual CO_2 emissions levels of the real and synthetic Guangdong in the years following the implementation of LCPI. Placebo studies and robustness tests were then performed to confirm the accuracy and robustness of the estimations.

3. Results

3.1. Real and Synthetic Guangdong

3.1.1. Total Effect

The Synth package for STATA provided by Abadie was used for the operation. Table 2 shows the weights of each province in the control group for synthesizing a virtual Guangdong. The results in Table 2 reveal that the carbon emissions trends in Guangdong before the implementation of LCPI are best copied by a combination of Hebei, Heilongjiang and Anhui. All other provinces in the control group were assigned zero W-weights.

Table 2. Province weights in synthetic Guangdong.

Province	Weight	Province	Weight
Hebei	0.630	Shandong	0
Shanxi	0	Henan	0
Inner Mongolia	0	Hunan	0
Jilin	0	Guangxi	0
Heilongjiang	0.255	Sichuan	0
Jiangsu	0	Guizhou	0
Anhui	0.115	Gansu	0
Fujian	0	Qinghai	0
Jiangxi	0	Ningxia	0

Table 3 compares the pre-treatment characteristics of real Guangdong with those of synthetic Guangdong as well as the average values of the 18 provinces in the control group. Based on the computed average, the other provinces were deemed unsuitable for the control group. Meanwhile, the carbon emission prediction factors of synthetic Guangdong were very similar to those of real Guangdong.

Table 3. Mean values of carbon emission predictors.

Variables	Real Guangdong	Synthetic Guangdong	Average of 18 Control Provinces
LnGDP	9.707	9.770	8.153
ES	51.14	51.01	46.87
LnPGDP	9.853	9.305	9.125
LnPOP	9.073	8.681	8.235
LnEC	9.523	9.356	8.687
LnTD	11.737	11.386	11.109
LnVN	14.816	14.028	13.281
CE-2000	199.6	193.1	105.3
CE-2005	341.8	347.7	205.9
CE-2009	437.6	444.0	283.7

Figure 4 presents the CO_2 emissions of the real and synthetic Guangdong from 1997 to 2015. Synthetic Guangdong tracks the changing trajectory of annual CO_2 emissions in real Guangdong very closely for the entire pre-LCPI period, thereby suggesting that synthetic Guangdong provides a sensible approximation of the annual CO_2 emissions in real Guangdong between 2010 and 2015 before the implementation of LCPI.

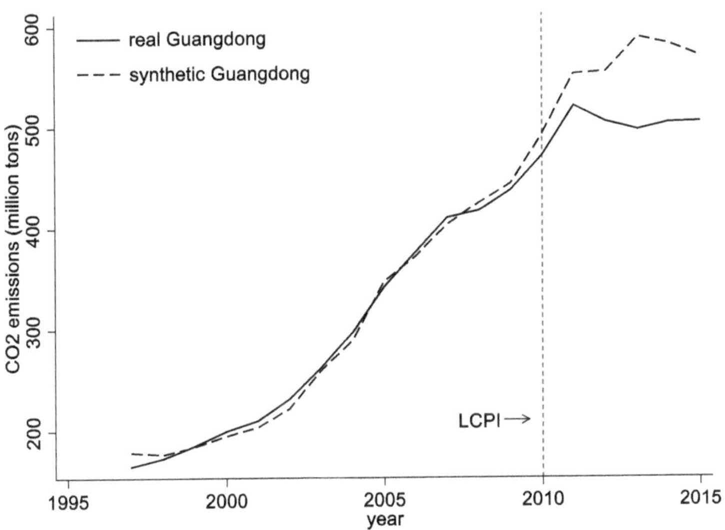

Figure 4. Trends in annual CO_2 emissions: Real vs. synthetic Guangdong.

The effect of LCPI on the carbon emissions of Guangdong was estimated as the difference between the annual CO_2 emissions of the real and synthetic version after the implementation of the policy. As shown in Figure 4, after the implementation of LCPI, the two lines began to diverge. Whilst the CO_2 emissions of synthetic Guangdong continued to grow steadily, those of real Guangdong began to shift after reaching its peak in 2011 and then declined for the first time. This figure suggests that the implementation of LCPI has a negative effect on Guangdong's carbon emissions. Table 4 shows the yearly gaps in the CO_2 emissions of the real and synthetic Guangdong, which can reflect the yearly estimated impacts of LCPI. The gap in the annual CO_2 emissions of real and synthetic Guangdong was maintained between 4 mt and 8 mt in the five years before the implementation of LCPI. However, after the implementation of this policy in 2010, this gap instantly expanded to 21.2 mt and increased year by year until reaching its maximum in 2013 (91.8 mt). These trends suggest that LCPI had a significant effect on carbon emissions, with the carbon emissions of Guangdong decreasing by approximately 10% during the entire 2010–2015 period with an average annual reduction of approximately 56 mt.

Table 4. Annual CO_2 emissions between 2005 and 2015: Real vs. synthetic Guangdong (mt).

Year	Real Guangdong	Synthetic Guangdong	Gap
2005	341.8	347.1	−5.3
2006	376.2	372.0	4.2
2007	410.1	403.3	6.8
2008	416.9	424.9	−8.0
2009	437.6	444.2	−6.6
2010	471.5	492.7	−21.2
2011	520.6	552.4	−31.8
2012	504.7	554.7	−50
2013	496.8	588.6	−91.8
2014	503.8	581.9	−78.1
2015	504.8	569.9	−65

3.1.2. Effects on Different Sources

Chapter 2 identifies energy production and manufacturing industries as the main sources of carbon emissions in Guangdong and suggests that LCPI mainly focuses on these two aspects. Therefore, what is the effect of LCPI on these two main sources of carbon emissions?

Figure 5 shows the results obtained by the synthetic control method, with Figure 5a showing the impact of LCPI on carbon emissions from energy production and Figure 5b showing the impact of LCPI on carbon emissions from manufacturing industries. These figures suggest that if Guangdong did not implement LCPI in 2010, the development trend of CO_2 emissions from energy production in synthetic Guangdong would become similar to those in real Guangdong, that is, the CO_2 emissions began to decline after reaching the peak in 2011. However, the role of LCPI in reducing carbon emissions from manufacturing industries in Guangdong is very obvious. After the implementation of this policy, the two lines began to diverge noticeably. Whilst the CO_2 emissions from the manufacturing industries of synthetic Guangdong continued to grow steadily, those of real Guangdong began to shift before declining. In this case, LCPI mainly affected the carbon emissions from the manufacturing industries of Guangdong yet had minimal effects on the carbon emissions from energy production.

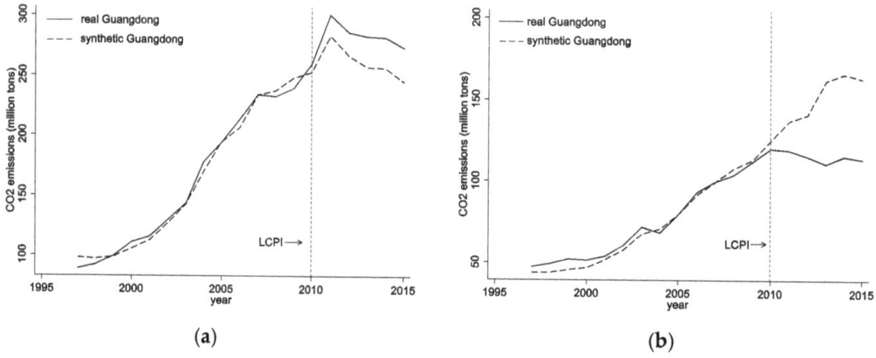

Figure 5. Trends in annual CO_2 emissions from different sources: Real vs. synthetic Guangdong. (**a**) Annual CO_2 emissions from energy production, and (**b**) annual CO_2 emissions from manufacturing industries.

3.2. Placebo Studies

Placebo studies were performed to verify whether the difference in the carbon emissions of real and synthetic Guangdong is indeed caused by the implementation of LCPI instead of other factors, that is, whether the empirical estimates are statistically significant. In these studies, Guangdong was transferred to the control group and a province from the existing control group was assumed to have implemented LCPI in 2010 instead of Guangdong. The synthetic control method was then employed to construct a synthetic version of the selected province and to estimate the effects of LCPI. All provinces in the control group were iteratively tested, and then the carbon emission differences generated in the placebo studies were compared with those obtained from the empirical analysis. If the differences in carbon emissions from Guangdong in the empirical analysis are much larger than those obtained in the placebo studies, then the difference in carbon emissions between real and synthetic Guangdong is driven by the implementation of LCPI, that is, the results of the empirical analysis are credible.

If the synthetic province has a large RMSPE (root-mean-square prediction error) before the implementation of LCPI, that is, the real carbon emission situation of this province before the LCPI implementation cannot be easily tracked, then the differences in the carbon emissions of the real and synthetic provinces post-LCPI are inferred to be caused by factors other than LCPI. Therefore, these provinces were eliminated in the placebo studies. Following Abadie (2015), those provinces whose pre-LCPI RMSPE was more than three times larger than that of Guangdong were excluded from the placebo studies. These provinces included Hebei, Inner Mongolia, Jiangsu, Shandong, and Qinghai [31].

The results of the placebo studies are presented in Figure 6, where the black line denotes the differences estimated for Guangdong and the grey line denoting the gap in annual CO_2 emissions between the provinces in the control group and their respective synthetic versions. The estimated gap in the carbon emissions of Guangdong from 2010 to 2015 was unusually large compared with those of the provinces in the control group. Given that the placebo studies included 14 provinces, the probability of the estimation result was computed as 1/14 = 0.071, which is statistically significant.

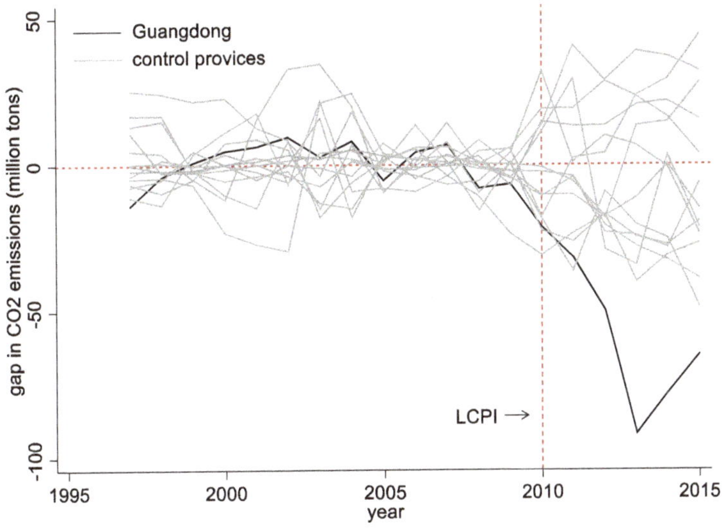

Figure 6. Annual CO_2 emissions gaps in Guangdong and placebo gaps in the 19 control provinces (excluding those provinces with a pre-low-carbon pilot initiative (LCPI) RMSPE that is three times larger than that of Guangdong).

3.3. Robustness Test

A robustness test was also performed based on an iterative approach, including the iterations of regions and the iterations of predictors. In the iterations of regions, a version of Guangdong was iteratively reconstructed, and each province that received a positive weight to the synthetic Guangdong, including Hebei (0.630), Heilongjiang (0.255), and Anhui (0.115), was deleted. This test aims to check whether the impact of LCPI on carbon emissions is influenced by weights and whether the results will differ when a province is removed from the control group. In the iterations of predictors, a version of Guangdong was iteratively reconstructed and a predictor was deleted at each iteration. This test aims to check whether the results will differ when a predictor is removed.

Figure 7 presents the trajectory of the annual CO_2 emissions of real Guangdong (solid line), synthetic Guangdong (long dashed line) and leave-one-out estimates (short dashed lines). Regardless of which iterative method was employed, the above figure shows that after removing the province with the positive weight from the control group or removing a predictor from the variable group, the CO_2 emissions of synthetic Guangdong (leave-one-out) remained much higher than those of real Guangdong, thereby suggesting that the results for the effects of LCPI on reducing the carbon emissions of Guangdong do not vary across the provinces in the control group or the changes in the predictors. In sum, the results of the previous analysis are fairly robust.

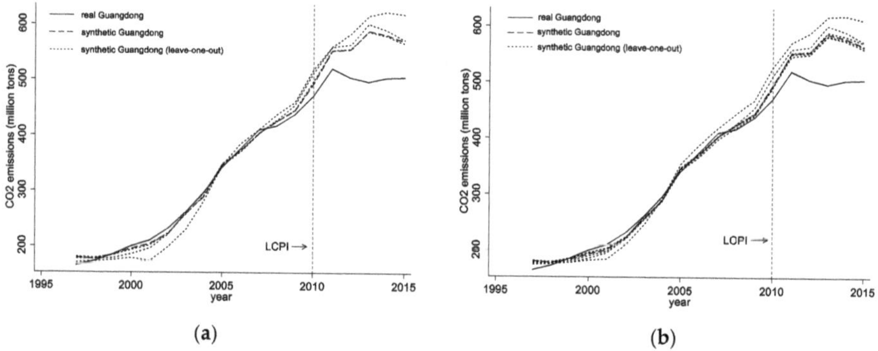

Figure 7. Leave-one-out distribution of the synthetic control for Guangdong. (**a**) Results based on the iterations of regions, and (**b**) results based on the iterations of predictors.

4. Discussion

As the world's top energy consumer and carbon emitter, China's carbon emissions policies, including LCPI, have important impacts on global climate change. Accurately assessing the effectiveness of these policies has important policy implications for China and the world in their pursuit to address global warming. This article selected Guangdong Province as the study area given that this province has the largest economic scale, population size and carbon emissions amongst China's low-carbon pilot provinces. The synthetic control method was applied to determine whether the implementation of LCPI reduced the carbon emissions of Guangdong. For the entire 2010–2015 period, the carbon emissions of Guangdong were reduced by approximately 10% as a result of the implementation of LCPI in July 2010. This policy had the greatest impact on the carbon emissions of manufacturing industries but had little impact on carbon emissions from energy production. Meanwhile, the placebo studies and robustness test confirmed that the findings are fairly robust.

This work differs from the previous literature in three ways. Firstly, to make the results closer to objective facts, instead of estimating carbon emissions, more uniform and authoritative carbon emissions data at the provincial level were collected from CEADs. Data-driven procedures instead of subjective selection were also employed to construct suitable comparison groups based on the synthetic control method. Secondly, the impact of low-carbon pilots on carbon emissions from different sources in Guangdong, including energy production and manufacturing industries, was analyzed. Thirdly, the findings of this work reveal that the impact of LCPI on China's regional carbon emissions does not increase every year. The impact of this policy on Guangdong's carbon emissions reached its maximum in 2013, after which Guangdong's carbon emissions began to rise and the impact of low-carbon pilots began to decline.

These results also offer several policy implications. Firstly, policy support and low-carbon action are necessary for reducing regional carbon emissions. Given that carbon is mainly produced via the consumption of fossil energy, a low-carbon development model characterized by low energy consumption, low pollution and low emissions must be developed to fundamentally solve the problem of global warming. The transformation of this model is inseparable from the power of the government, especially for China, which has a high concentration of administrative power.

Secondly, low-carbon policies must be constantly adjusted during their implementation. These policies must not be static and must keep pace with the times. The government must adjust the content of its low-carbon policies based on feedback in order to continuously optimize such policies and ensure their contributions to reducing regional carbon emissions.

Thirdly, the successful low-carbon action experiences in low-carbon pilot areas are worth learning and promoting in other regions. The significance of LCPI lies not only in its goal to achieve a quantitative emission reduction but also in its exploration of low-carbon development models through pilots, its search for effective and sustainable emission reduction methods and the accumulation of relevant experiences that can be popularized on a larger scale. Since the implementation of LCPI, the low-carbon pilot areas have accumulated many useful experiences in low-carbon development through practice. For example, Guangdong explored a low-carbon development model with low-carbon industry and low-carbon technologies as the core, low-carbon energy as the support and low-carbon energy, low-carbon transportation, low-carbon buildings and low-carbon lifestyle as the base. A series of auxiliary systems, including a total carbon emission control system, inclusive low-carbon system and carbon emissions trading system, was also implemented in the province to reduce its carbon emissions. Guangdong also launched the 'nine plus two' carbon cooperation mechanism in the Pan-Pearl River Delta region to transform itself into the first low-carbon pilot province in China. These practices can greatly help the other regions of China in establishing a low-carbon economy.

Social scientists are always looking for ways to accurately evaluate the effectiveness of policies. This paper uses a quantitative analysis model, namely, the synthetic control method, to evaluate the impact of low-carbon pilot initiatives on regional carbon emissions as well as builds its comparison groups through data drivers instead of subjective selection to enhance the objectivity and accuracy of the estimations. However, the synthetic control method has several limitations. For instance, when constructing a synthetic object, all objects in the control group must have a positive weight and sum of 1. Therefore, the restrictions on weights must be relaxed and negative weights must be incorporated in future attempts to improve such method. In addition, LCPI has been implemented for nearly nine years, and future studies may present a summarization of experiences and popularize such policy to achieve a comprehensive low-carbon development.

Author Contributions: Conceptualization, X.Y. and M.S.; data collection, X.Y. and D.W.; analyzed data, X.Y. and D.W.; writing—review and editing, X.Y., D.W. and B.T.I.

Funding: This research was funded by the National Social Science Fund of China, grant number (No. 16ZDA050), the National Natural Science Foundation of China (No. 71874092).

Conflicts of Interest: The authors declare no conflict of interest.

References

1. Anthes, R.A.; Corell, R.W.; Holland, G.; Hurrell, J.W.; MacCracken, M.C.; Trenberth, K.E. Hurricanes and global warming—Potential linkages and consequences. *Bull. Am. Meteorol. Soc.* **2006**, *87*, 623–628. [CrossRef]

2. Hughes, T.P.; Kerry, J.T.; Baird, A.H.; Connolly, S.R.; Dietzel, A.; Eakin, C.M.; Heron, S.F.; Hoey, A.S.; Hoogenboom, M.O.; Liu, G.; et al. Global warming transforms coral reef assemblages. *Nature* **2018**, *556*, 492. [CrossRef] [PubMed]

3. Raper, S.C.B.; Braithwaite, R.J. Low sea level rise projections from mountain glaciers and icecaps under global warming. *Nature* **2006**, *439*, 311–313. [CrossRef] [PubMed]

4. Li, Y.; Xie, Z.; Qin, Y.; Xia, H.; Zheng, Z.; Zhang, L.; Pan, Z.; Liu, Z. Drought Under Global Warming and Climate Change: An Empirical Study of the Loess Plateau. *Sustainability* **2019**, *11*, 1281. [CrossRef]

5. Matthews, H.D.; Gillett, N.P.; Stott, P.A.; Zickfeld, K. The proportionality of global warming to cumulative carbon emissions. *Nature* **2009**, *459*, 829. [CrossRef] [PubMed]

6. Allen, M.R.; Frame, D.J.; Huntingford, C.; Jones, C.D.; Lowe, J.A.; Meinshausen, M.; Meinshausen, N. Warming caused by cumulative carbon emissions towards the trillionth tonne. *Nature* **2009**, *458*, 1163–1166. [CrossRef] [PubMed]

7. Lashof, D.A.; Ahuja, D.R. Relative contributions of greenhouse gas emissions to global warming. *Nature* **1990**, *344*, 529–531. [CrossRef]

8. Knight, K.W.; Schor, J.B. Economic Growth and Climate Change: A Cross-National Analysis of Territorial and Consumption-Based Carbon Emissions in High-Income Countries. *Sustainability* **2014**, *6*, 3722–3731. [CrossRef]

9. Hamers, L. Global Carbon Dioxide Emissions Will Hit a Record High in 2018. Available online: https://www.sciencenews.org/article/global-carbon-dioxide-emissions-will-hit-record-high-2018 (accessed on 5 March 2019).

10. Le Quéré, C.; Andrew, R.M.; Friedlingstein, P.; Sitch, S.; Hauck, J.; Pongratz, J.; Pickers, P.A.; Korsbakken, J.I.; Peters, G.P.; Canadell, J.G.; et al. Global Carbon Budget 2018. *Earth Syst. Sci. Data* **2018**, *10*, 2141–2194.

11. Shan, Y.L.; Guan, D.B.; Zheng, H.R.; Ou, J.M.; Li, Y.; Meng, J.; Mi, Z.F.; Liu, Z.; Zhang, Q. Data Descriptor: China CO_2 emission accounts 1997–2015. *Sci. Data* **2018**, *5*, 170201. [CrossRef]

12. NBS. *China Statistical Yearbook*; China Statistics Press: Beijing, China, 2018.

13. Fankhauser, S.; Kennedy, D.; Skea, J. Building a low-carbon economy: The inaugural report of the UK Committee on Climate Change. *Glob. Environ. Chang. Part B Environ. Hazards* **2009**, *8*, 201–208. [CrossRef]

14. Chen, W.Y.; Yin, X.; Zhang, H.J. Towards low carbon development in China: A comparison of national and global models. *Clim. Chang.* **2016**, *136*, 95–108. [CrossRef]

15. Winkler, H.; Marquand, A. Changing development paths: From an energy-intensive to low-carbon economy in South Africa. *Clim. Dev.* **2009**, *1*, 47–65. [CrossRef]

16. Schwanen, T. The Bumpy Road toward Low-Energy Urban Mobility: Case Studies from Two UK Cities. *Sustainability* **2015**, *7*, 7086–7111. [CrossRef]

17. NDRC. Notice on ISSUANCE Medium and Long-Term Special Plan for Energy Conservation. Available online: http://www.ndrc.gov.cn/fzggzy/hjbh/jnjs/200507/t20050711_45823.html (accessed on 18 February 2019).

18. State Council of China. Notice on the Issuance of China's National Climate Change Program. Available online: http://www.gov.cn/gongbao/content/2007/content_678918.htm (accessed on 25 December 2018).

19. State Council of China. China's Policies and Actions for Addressing Climate Change. Available online: http://www.ccchina.gov.cn/WebSite/CCChina/UpFile/File419.pdf (accessed on 25 December 2018).

20. State Council of China. China to Cut 40 to 45% GDP Unit Carbon by 2020. Available online: http://www.chinadaily.com.cn/china/2009-11/26/content_9058731.htm (accessed on 25 December 2018).

21. National People's Congress. The Twelfth Five-Year Plan on National Economy and Social Development. Available online: http://news.sina.com.cn/c/2011-03-17/055622129864.shtml (accessed on 10 November 2018).

22. Chu, G.D.; Gao, Z.Y. Empirical analysis of the effect of in low-carbon-city pilots. *Co-Oper. Econ. Sci.* **2017**, *20*, 4–7.

23. Deng, L.L.; Zhan, J. Does two low carbon pilot policy promote the performance of carbon emissions reduction in pilot cities—Based on difference in difference method. *Syst. Eng.* **2017**, *35*, 68–73.

24. Dai, R.; Chao, J.H. The effect of China's first low-carbon pilot: DID estimation of five cities and eight provinces. *Sci. Technol. Manag. Res.* **2015**, *35*, 56–61.

25. Lu, X.W. Study on the effectiveness of the low carbon policy: Evidence from the synthetic control methods. *Soft Sci.* **2017**, *31*, 98–101, 109.

26. Zhao, S.; Chen, R.; Jiang, Z.D. Empirical analysis of the low-carbon pilot policies' impact on agricultural carbon emissions based on DID model. *Ecol. Econ.* **2018**, *34*, 22–28.

27. Shan, Y.L.; Liu, J.H.; Liu, Z.; Xu, X.W.H.; Shao, S.; Wang, P.; Guan, D.B. New provincial CO_2 emission inventories in China based on apparent energy consumption data and updated emission factors. *Appl. Energy* **2016**, *184*, 742–750. [CrossRef]
28. Abadie, A.; Diamond, A.; Hainmueller, J. Synthetic Control Methods for Comparative Case Studies: Estimating the Effect of California's Tobacco Control Program. *J. Am. Stat. Assoc.* **2010**, *105*, 493–505. [CrossRef]
29. Lo, K.; Li, H.; Chen, K. Climate experimentation and the limits of top-down control: Local variation of climate pilots in China. *J. Environ. Plan. Manag.* **2019**, 1–18. [CrossRef]
30. Wang, F.; Wu, L.H.; Yang, C. Driving factors for growth of carbon dioxide emissions during economic development in China. *Econ. Res. J.* **2010**, *45*, 123–136.
31. Abadie, A.; Diamond, A.; Hainmueller, J. Comparative Politics and the Synthetic Control Method. *Am J. Polit. Sci.* **2015**, *59*, 495–510. [CrossRef]

Article

Research on the Spatial Network Characteristics and Synergetic Abatement Effect of the Carbon Emissions in Beijing–Tianjin–Hebei Urban Agglomeration

Xintao Li [1], Dong Feng [1,*], Jian Li [1,2] and Zaisheng Zhang [1]

[1]   College of Management and Economics, Tianjin University, Tianjin 300072, China;
      t391753064@tju.edu.cn (X.L.); lijian631219@163.com (J.L.); zhangzs@tju.edu.cn (Z.Z.)
[2]   Research Center for Circular Economy and Enterprise Sustainable Development, Tianjin University of
      Technology, Tianjin 300384, China
*   Correspondence: fengdong@tju.edu.cn

Received: 23 January 2019; Accepted: 5 March 2019; Published: 8 March 2019

Abstract: Based on the carbon emission data in the Beijing–Tianjin–Hebei urban agglomeration from 2007 to 2016, this paper used the method of social network analysis (SNA) to investigate the spatial correlation network structure of the carbon emission. Then, by constructing the synergetic abatement effect model, we calculated the synergetic abatement effect in the cities and we empirically examined the influence of the spatial network characteristics on the synergetic abatement effect. The results show that the network density first increased from 0.205 in 2007 to 0.263 in 2014 and then decreased to 0.205 in 2016; the network hierarchy fluctuated around 0.710, and the minimum value of the network efficiency was 0.561, which indicates that the network hierarchy structure is stern and the network has good stability. Beijing and Tianjin are in the center of the carbon emission spatial network and play important "intermediary" and "bridge" roles that can have better control over other carbon emission spatial spillover relations between the cities, thus the spatial network of carbon emissions presents a typical "center–periphery" structure. The synergetic abatement effect increased from −2.449 in 2007 to 0.800 in 2011 and then decreased to −1.653 in 2016; the average synergetic effect was −0.550. This means that the overall synergetic level has a lot of room to grow. The carbon emission spatial network has a significant influence on the synergetic abatement effect, while increasing the network density and the network hierarchy. Decreasing the network efficiency will significantly enhance the synergetic abatement effect.

Keywords: carbon emissions; spatial network; synergetic abatement; Beijing–Tianjin–Hebei urban agglomeration; SNA

---

1. Introduction

With the carbon dioxide concentration in the atmosphere increasing, a series of problems, such as global warming, glacier melting, sea level rise, and so on, have been triggered, which pose a serious threat to the sustainable development of human society and the security of the ecology and the environment [1]. The signing of the Paris Agreement demonstrates the determination of the international community to actively cope with global climate change. The signatories agreed to control the average rise of the global temperature within less than 2 °C above the pre-industrial level and to strive to control the temperature rise within 1.5 °C above the pre-industrial level. However, the annual Greenhouse Gas Bulletin issued by the World Meteorological Organization in 2017 pointed out that the average level of carbon dioxide in 2016 was 145% of that of pre-industrialization, and the average concentration of carbon dioxide in the world reached 403.3 parts per million, which surpassed the historical record [2]. This shows that working to reduce greenhouse gas emissions, mainly composed of carbon dioxide, and promote the coordinated development of the socio-economic and ecological environment are still major issues that need to be solved urgently in all countries of the world.

As the largest developing country in the world, China has become one of the largest carbon emitters [3,4]. How to deal with this contradiction between development and the environment in China has always been a global focus [5]. In order to solve this problem, the government of China is committed to achieving the goal of cutting carbon-emitting intensity by 40–45% before 2020 and plans to reach the peak of carbon emissions around 2030. Because of the vast territory of China and the differences in the development level of each region, natural resources endowment, and geographical environment [6], it is difficult for all regions to achieve the low-carbon development goals formulated by the state in the same way [7]. Therefore, finding a coordinated development path in terms of economy, society, and the ecological environment that is suitable for China's national conditions is the major premise for achieving the goal of carbon emission reduction. In fact, China is currently implementing a series of regional and spatial development strategies, such as the synergetic development of Beijing, Tianjin, and Hebei and the construction of the Yangtze River Economic Zone [8]. The spatial correlation of carbon emissions has broken through the traditional linear model and is gradually presenting a nonlinear spatial structure with complex network structure characteristics. Thus, in order to adhere to green and low-carbon development, it is an important policy for China to establish a new mechanism of regionally coordinated development with urban agglomerations as the main body [9]. Against this background, this paper chose the Beijing–Tianjin–Hebei urban agglomeration as the research area, focusing on the spatial network characteristics of carbon dioxide emissions and the synergetic emission reduction effect. Based on the relational data and network analysis perspective, this took the spatial correlation of carbon emissions as the breakthrough point, analyzing the spatial network structure and the effect of its carbon emissions with the help of the social network analysis method. This aimed at formulating emission reduction policies that are more in line with regional characteristics in the implementation of a regional coordinated development strategy in China, as well as providing a reference and basis for achieving synergetic emission reduction among regional cities.

2. Literature Review

Carbon emissions have always been a hot issue in academic research and also a key issue in regional socio-economic development. Early studies focused on the relationship between carbon emissions and economic development [10,11] and systematically explored the principles and mechanisms of carbon emission reduction [12,13]. On this basis, scholars at home and abroad have made great research achievements regarding carbon-emission intensity [14], carbon emission performance [15], carbon emission quota [16], and factors affecting carbon emissions [17]. According to recent studies, the spatial effects of carbon emissions have become an important research subject. Many scholars use exploratory spatial data analysis (ESDA) to examine the characteristics of the spatial correlation and spatial agglomeration of carbon emissions at the regional level. Grunewald et al. explored the driving factors for the formation of spatial differences in carbon emissions and found that energy intensity and energy structure were the main reasons for these spatial differences [18]. Marbuah and Mensah took 290 urban districts in Sweden as research areas. Statistical tests were carried out to analyze the spatial correlation of several pollutants, including carbon dioxide. The results showed that the spatial spillover effect was the main driving factor of the environmental Kuznets curve [19]. Through spatial econometrics, Wu studied the temporal and spatial pattern and evolution mechanism of carbon emission reduction in different provinces of China and analyzed the emission reduction characteristics of key provinces [20]. Wang et al. constructed spatial econometric models, which were based on patent data about energy conservation and emission reduction and economic externalities of spatial units that were transplanted into CO_2 emissions research [21]. Yan et al. used the undesirable-slacks-based measure (SBM) model and evaluated the carbon emission efficiency [22]. You et al. applied the spatial panel method to address the problems of spatial dependency and the spillover effect among neighboring countries [23]. Sun et al. divided China into resource-based and non-resource-based areas according to their industrial development level. Using the equity measurement method, the dynamic differences of regional carbon emissions were discussed and the causes of profit-loss deviation were revealed [24]. Wang et al. estimated the spatial dependence of the carbon emission reduction potential at the provincial level, combined with the economic weight matrix and the novel spatial panel data model [25]. Zhang et al. used Arc GIS to calculate the carbon emission and absorption rates in China's Beijing–Tianjin–Hebei agglomeration [26]. From the perspective of research methods, some studies used social network analysis (SNA) [27–29] to analyze the regional carbon emission spatial network structure. Most of the research on carbon emissions focuses on carbon emission measurement and decomposition, carbon emissions' influencing factors, and carbon emission efficiency [30–32]. When using traditional spatial measurement methods to analyze carbon emissions and their influencing factors, most of the literature only considers geographically similar factors. An increasing number of researchers have proved that carbon emissions do not exist independently in various regions but in a certain spatial correlation. Existing research often considers the spatial correlation of carbon emissions from the perspective of provinces or regions and from the empirical study of "attribute data" of carbon emissions. They do not reveal the linkage structure between regions from the perspective of "relational data". In addition, there are a few issues involved with the synergy effect of carbon emission reduction among regions and how to strengthen the synergy of regional emission reduction.

In summary, previous studies have thoroughly explored the issue of carbon emissions and revealed the spatial correlation characteristics and spatial differences of carbon emissions to a certain extent, which has laid the foundation for the study of regional synergetic carbon emission reduction. However, the "relational data" have more analytical value, as they often determine the performance of the "attribute data". Therefore, on the basis of existing research, this paper uses the social network analysis method to comprehensively deconstruct the spatial network characteristics of carbon emissions in the urban agglomeration of Beijing, Tianjin, and Hebei, to analyze its synergetic emission reduction effect. On this basis, the proposal of synergetic carbon emission reduction in an urban agglomeration is put forward in order to understand the spatial correlation of carbon emissions as a whole and to reveal the interconnected network structure of carbon emissions among regions and its impact on synergetic carbon emission reduction. This can provide a decision-making reference for achieving the goal of coordinated regional carbon emission reduction and formulating corresponding carbon emission reduction policies.

3. Research Methods and Sources of Data

3.1. Construction of a Spatially Relevant Network of Carbon Emissions

Social network analysis is a quantitative analysis method for relational data, which has been widely used in sociology, economics, management, and other disciplines. It takes "relationship" as the basic unit of analysis and expresses the interaction among members as a pattern or rule based on the relationship by studying the network relationship. The confirmation of relationships is the key to the analysis of social networks. Currently, the methods of describing spatial correlation are mainly the Granger Causality test based on the VAR model and the gravitational model or the improved gravitational model [33–36]. This paper uses the improved gravity model to construct the spatial correlation network of carbon emissions in the urban agglomeration of Beijing–Tianjin–Hebei, as the VAR model is too sensitive to the selection of lag order, which would reduce the accuracy of the characterization of the network structure characteristics [37–39]. In addition, the gravitational model is not only applicable to the total data, but also can take economic and geographical factors into account. The basic model is:

$$G_{ij} = \frac{C_i}{C_i + C_j} \times \frac{\sqrt[3]{P_i C_i E_i} \sqrt[3]{P_j C_j E_j}}{\left(\frac{D_{ij}}{e_i - e_j}\right)^2}. \tag{1}$$

In Formula (1), i and j indicate i city and j city in the urban agglomeration of Beijing–Tianjin–Hebei respectively; G_{ij} indicates the carbon emission gravity between cities; C_i and C_j indicate the carbon emission of i city and j city, respectively; P_i and P_j indicate the total population of i city and j city, respectively; E_i and E_j indicate the gross product of i city and j city, respectively; D_{ij} indicates the spatial distance between i city and j city; e_i and e_j indicate the per capita production of i city and j city, respectively. $\frac{D_{ij}}{e_i - e_j}$ Value means "economic distance" between i city and j city.

According to Formula (1), the gravitational matrix of carbon emissions of the urban agglomeration of Beijing–Tianjin–Hebei can be calculated, then the average value of each row of the gravitational matrix can be taken as the critical value. If the gravity is higher than the average value, then the value is taken as 1, which indicates that there is a correlation between the city in this row and the carbon emission of the city in this column. In contrast, if the gravity is lower than the average value, the value is taken as 0, indicating that there is no correlation between the two. Thus, the spatial correlation matrix of carbon emissions in the urban agglomeration of Beijing–Tianjin–Hebei can be constructed.

3.2. Characteristic Index of the Spatial Network of Carbon Emissions

In social network analysis, four indicators—network density, network correlation degree, network hierarchy, and network efficiency—are commonly used to reflect the overall characteristics of the network structure, while for the individual network structure characteristics, degree centrality and intermediary centrality are used to describe them [40,41].

For the structural characteristics of the entire network, if N is the number of urban nodes in the carbon emission network of the urban agglomeration of Beijing–Tianjin–Hebei, the largest number of possible correlations in the network is $N \times (N-1)$; if the number of actual correlations in the network is M, the density of the network can be expressed as follows:

$$Dn = M/[N \times (N-1)]. \tag{2}$$

Network density reflects the closeness of the entire network structure. The greater the network density, the closer the relationship between the carbon emissions of the cities in Beijing, Tianjin, and Hebei and the greater the impact of the network structure on the carbon emissions of the cities.

If the number of pairs of unreachable points in the network is V, the formula for calculating the degree of network correlation is as follows:

$$C = 1 - V/[N \times (N-1)/2]. \tag{3}$$

The network correlation degree reflects the robustness and vulnerability of the spatial network of carbon emissions. The greater the network correlation degree, the greater the direct or indirect correlation between the cities in the spatial network of carbon emissions of the urban agglomeration of Beijing–Tianjin–Hebei.

If the number of pairs of symmetrically reachable points in the network is K, $\max(K)$ is the number of pairs of the most possible symmetrically reachable points in the network. Then, the hierarchy of the network can be expressed as:

$$H = 1 - K/\max(K). \tag{4}$$

Network hierarchy describes the degree of asymmetric accessibility of cities in the network. The higher the network hierarchy, the stricter the hierarchical structure among the cities. Only a few cities are in the dominant position in the network.

If M is the number of redundant lines in the network, $\max(M)$ is the maximum possible number of redundant lines. The formula for calculating the network efficiency of the network is as follows:

$$E = 1 - M/\max(M). \tag{5}$$

Network efficiency characterizes the connective efficiency between cities in the network. The lower the network efficiency, the more links that exist between cities in the network. Therefore, their carbon emissions are closely related, generating more channels of spatial spillover.

For the structural characteristics of the individual network, if d_i is the number of direct correlations existing between i city and other cities in the carbon emission network of the urban agglomeration of Beijing–Tianjin–Hebei, then the degree of centrality of i city can be expressed as:

$$C_d(i) = d(i)/(N-1). \tag{6}$$

Point-degree centrality measures the degree to which the cities of Beijing, Tianjin, and Hebei are at the center of the spatial correlation network of carbon emissions. The greater the degree of point centrality, the more the city is at the center of the network and the more connections it has with other cities.

If g_{jk} is the number of relationship paths between j city and k city and $g_{jk}(i)$ is the number of cities to pass through the relationship paths between j city and k city, the formula for calculating the degree of intermediary centrality of cities is as follows:

$$C_b(i) = \frac{2\sum\limits_{j<k} \left[g_{jk}(i)/g_{jk}\right]}{(N-1)(N-2)}. \tag{7}$$

The degree of intermediary centrality reflects the extent to which a city in the network can control the relationship between the other cities. The greater the degree of intermediary centrality, the stronger the "intermediary" role the city plays in the network, thus the greater the extent to which it can control the correlation of carbon emissions of other cities [42].

3.3. Synergetic Effect Model of Carbon Emission Reduction

Due to externalities, the problem of environmental pollution needs to be treated through regional coordination in order to make fundamental achievements. In the process of regional coordinated emission reduction, regions with higher economic development levels and lower pollution levels are regarded as learning examples for other regions [43]. If other regions can change the direction and extent of pollutant emission in the same direction as the benchmark areas, the benchmark effect of coordinated emission reduction will exist among the regions [44,45]. Considering this, this paper chose Beijing, the central city of the urban agglomeration of Beijing–Tianjin–Hebei, as the benchmark area of carbon emission reduction and used the methods of Cerqueira and Martins to construct the synergetic effect model of carbon emission reduction of the urban agglomeration of Beijing–Tianjin–Hebei [46].

$$Corr_{bi,t} = 1 - \frac{1}{2}\left[\frac{(c_{b,t} - \overline{c_b})}{\sqrt{\frac{1}{t}\sum\limits_1^t (c_{b,t} - \overline{c_b})^2}} - \frac{(c_{i,t} - \overline{c_i})}{\sqrt{\frac{1}{t}\sum\limits_1^t (c_{i,t} - \overline{c_i})^2}}\right]^2 \tag{8}$$

In Formula (8), $Corr_{bi,t}$ shows the synergetic effect of carbon emission reduction of i city of the urban agglomeration of Beijing–Tianjin–Hebei for the year and the central city, Beijing. The closer the value is to 1, the higher the synergy between the i city and the central city, Beijing. $c_{b,t}$ indicates the total annual carbon emissions of Beijing for the t year and $\overline{c_b}$ is the average annual carbon emissions of Beijing for the t year. $c_{i,t}$ indicates the total annual carbon emissions of i city in the urban agglomeration of Beijing–Tianjin–Hebei for the t year and $\overline{c_i}$ shows the average annual carbon emissions of i city in the urban agglomeration of Beijing–Tianjin–Hebei.

3.4. Sources of Data

This paper takes 13 cities, including Beijing, Tianjin, Shijiazhuang, Chengde, Zhangjiakou, Qinhuangdao, Tangshan, Langfang, Baoding, Cangzhou, Hengshui, Xingtai, Handan, as the research objects. The research time ranges from 2007 to 2016—10 years in total. The data of economy, population, and energy consumption of each city come from the statistical annals of Chinese cities, the statistical annals of China's energy resources, the statistical annals of Beijing, Tianjin, and Hebei, and the statistical annals and bulletins of Hebei cities in each year; these data are properly referred to.

Considering the lack of statistical data on carbon dioxide emissions at the urban level, this paper adopts the simplified estimation method of carbon dioxide emissions based on apparent energy consumption in the 2006 IPCC Guidelines for National Greenhouse Gas Inventories to calculate the carbon dioxide emissions of each city in the urban agglomeration of Beijing–Tianjin–Hebei. The calculation formula is as follows:

$$E_{co_2} = \sum_i (E_i \times V_i \times F_i \times O_i) \times \frac{44}{12},$$

(9)

where E_{co_2} is the total amount of carbon dioxide emissions generated by energy consumption, E_i is the apparent consumption of the i energy, V_i is the low calorific value of the i energy, F_i is the carbon emission factor of the i energy, and O_i is the oxidation rate of the energy. The conversion coefficient of $\frac{44}{12}$ converts the carbon atom mass to the carbon dioxide molecular mass. Considering the availability of the data, this paper chooses several kinds of energy sources—coal, coke, crude oil, gasoline, kerosene, diesel, raw oil, liquefied petroleum gas, and natural gas—to calculate the carbon dioxide emissions of each city in the urban agglomeration of Beijing–Tianjin–Hebei [47–50]. Among them, the correlation coefficients of various energy sources are mainly derived from the data published in the 2006 IPCC Guidelines for National Greenhouse Gas Emissions Inventories.

4. Calculation Results and Analysis

4.1. Characteristic Analysis of the Spatial Network of Carbon Emissions

According to the improved gravity model of Formula (1) mentioned above, the carbon emission gravity matrix of the urban agglomeration of Beijing–Tianjin–Hebei is primarily calculated and converted into a spatial correlation matrix. In order to more effectively demonstrate the structure of the spatial correlation network of carbon emissions of the urban agglomeration of Beijing–Tianjin–Hebei, this paper draws a spatial correlation network of carbon emissions in 2016 by using the visualization tool Net-draw from the software UCINET6.0 (UCINET v6.0, Analytic Technologies, Lexington, KY, USA). The specific results of the network are shown in Figure 1.

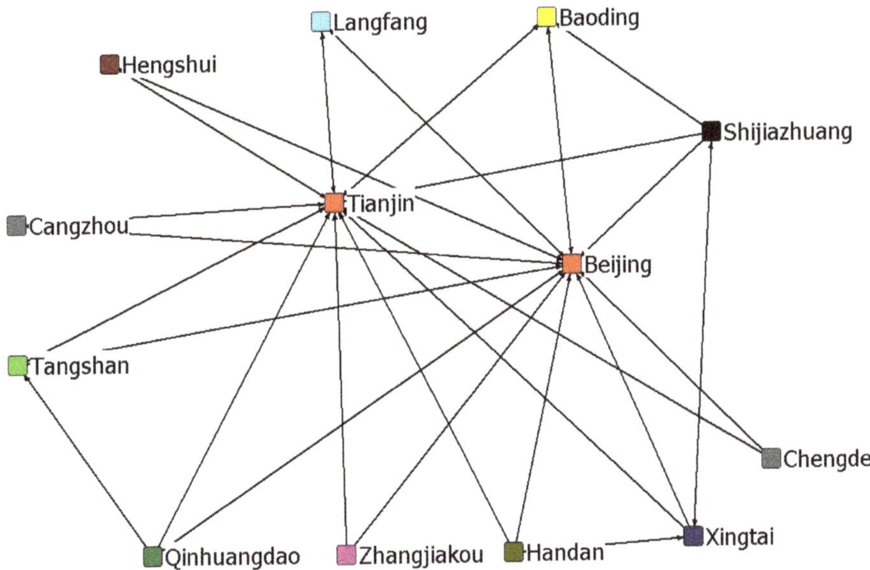

Figure 1. Spatially relevant network of carbon emissions of the urban agglomeration of Beijing–Tianjin–Hebei in 2016.

As is shown in Figure 1, the spatial correlation network of carbon emissions in the urban agglomeration of Beijing–Tianjin–Hebei takes on a typical "center–periphery" structure. There are 32 correlations among the 13 cities and there is at least one spatial correlation among each city, which indicates that the carbon emissions of each city are generally related in space, and the urban agglomeration of Beijing–Tianjin–Hebei has a relatively stable spatial network structure of carbon emissions.

In order to further investigate the spatial network characteristics of carbon emissions in the urban agglomeration of Beijing–Tianjin–Hebei, the network density, network correlation, network hierarchy, network efficiency, point degree centrality, and intermediary centrality were calculated based on Formula (2–7). The spatial network characteristics of carbon emissions of the urban agglomeration of Beijing–Tianjin–Hebei from 2007 to 2016 were analyzed from two aspects: The entire network's characteristics and individual centrality. Firstly, the overall characteristics of the spatial network of carbon emissions in the urban agglomeration of Beijing–Tianjin–Hebei were calculated; the specific results are shown in Figure 2.

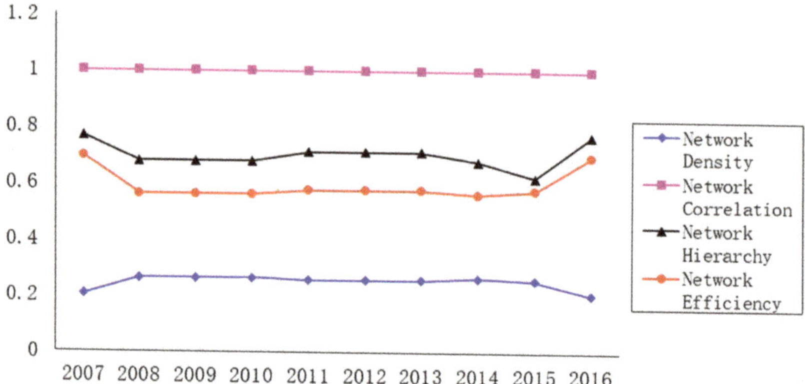

Figure 2. Overall characteristic indicators of the spatial network of carbon emissions in the urban agglomeration of Beijing–Tianjin–Hebei from 2007 to 2016.

According to the results of Figure 2, the network density of the spatial network of carbon emissions in the urban agglomeration of Beijing–Tianjin–Hebei increased first and then decreased from 0.205 in 2007 to 0.263 in 2014, then to 0.205 in 2016. The increase of network density shows that the carbon emission links between the cities of Beijing, Tianjin, and Hebei are getting closer, but the network density has declined since 2014, which indicates that the spatial links of carbon emissions in the urban agglomeration of Beijing–Tianjin–Hebei have been reduced and it is imperative to enhance the coordinated emission reduction among the cities. On the other hand, the correlation degree of the spatial network of carbon emissions in the urban agglomeration of Beijing–Tianjin–Hebei over the years is 1, which indicates that the structure of the carbon emission network of the urban agglomeration of Beijing–Tianjin–Hebei is stable, there is general spatial correlation and spillover, and the network accessibility is good. From 2007 to 2016, the network hierarchy fluctuated around 0.710, which indicates that, in the spatial network of carbon emissions, the hierarchical structure between the cities of Beijing, Tianjin, and Hebei is stricter and only a few cities are in the center of the network. The network efficiency shows an oscillatory trend, with a minimum value of 0.561 and a maximum value of 0.697. The value of the comprehensive network efficiency is relatively high, which indicates that there are more redundant connections in the network and there are many overlaps in the spatial spillover of carbon emissions. This also means that the existing network correlation can maintain the stability of the network. The results indicate that the carbon emission spatial correlation is increasingly close in the urban agglomeration. This result may have originated from China, which put forward the goal of energy conservation and emission reduction starting in 2006. The carbon emission spatial correlation was enhanced as the adjustment of the energy and industrial structure, the energy flow, and the industrial transfer among the cities were gradually increased.

To compare with the existing research [27–29] on China's provincial carbon emission spatial network, our analysis shows that in the process of China's current urban agglomeration development, there also exists a stable and complex carbon emission spatial network structure and correlation relationship among the cities in the urban agglomeration. The difference is that, in the national carbon emission spatial network, the network hierarchical structure is gradually broken and an increasing number of small provinces have changed the subordinate and edge positions in the network. However, as the results show in Figure 2, the large cities are still in the center position and have the control status. In other words, under the influence of the carbon emission spatial network, in order to improve the carbon emission reduction efficiency of the urban agglomeration, the small cities in Hebei province also need to take actions to change the subordinate and edge positions in the network. Only focusing on the emission reduction measures of the large cities, such as Beijing and Tianjin, will not produce a long-term mechanism for the realization of the regional emission reduction goals.

Furthermore, according to Formula (6,7), two indices of each city in the urban agglomeration of Beijing–Tianjin–Hebei—point degree centrality and intermediary degree—are calculated. Individual centrality of the carbon emission network in the urban agglomeration of Beijing–Tianjin–Hebei is analyzed but, due to the limitation of the length of the paper, only the calculation results of 2016 are given, as shown in Table 1.

Table 1. Individual centrality analysis of the spatial network of carbon emissions in the urban agglomeration of Beijing–Tianjin–Hebei in 2016.

City	Out Degree	In Degree	Degree Centrality (%)	Rank	Betweenness Centrality (%)	Rank
Beijing	2	9	91.667	1	7.071	2
Tianjin	3	9	91.667	1	15.404	1
Shijiazhuang	4	1	33.333	3	0.253	6
Chengde	2	0	16.667	9	0.000	8
Zhangjiakou	2	0	16.667	9	0.000	8
Qinhuangdao	3	0	25.000	5	0.000	8
Tangshan	2	1	25.000	5	0.000	8
Langfang	2	2	16.667	9	1.010	3
Baoding	2	3	25.000	5	1.010	3
Cangzhou	2	1	16.667	9	0.253	6
Hengshui	2	0	16.667	9	0.000	8
Xingtai	3	2	33.333	3	0.758	5
Handan	3	0	25.000	5	0.000	8

From the results of Table 1, we can see that only three of the 13 cities, Tianjin, Beijing, and Baoding, in the urban agglomeration of Beijing–Tianjin–Hebei have a higher point-in degree than point-out degree. They belong to the recipient subject in the spatial network correlation of carbon emissions, while the other ten cities have a higher point-out degree than point-in degree, belonging to the spillover subject in the spatial network correlation of carbon emissions. From the results of the point-degree centrality, the point-degree centrality of Beijing and Tianjin is up to 91.667, which is obviously higher than that of other cities. This shows that Beijing and Tianjin are in the central position in the carbon emission network and have more links with other cities. The cities with a relatively small degree of centrality tend to send out a carbon emission correlation to the cities with a larger degree of centrality. In terms of the intermediary centrality, Tianjin and Beijing still ranked in the first two places, with 15.404 and 7.071, respectively. The results of the other cities are far lower than those of Tianjin and Beijing. It can be seen that Beijing and Tianjin, which lie in the center of the network, also play an important role of "intermediary" and "bridge" in the spatial network of carbon emissions, while the other cities are in a weaker position in the network. This is mainly because, as the two core cities in the urban agglomeration of Beijing–Tianjin–Hebei and compared to the cities of Hebei Province, Beijing and Tianjin have benefited in terms of economic development, industrial structure, energy consumption, and so on. This reveals that there is a significant carbon emission spatial transfer trend from energy-rich cities to economically developed cities in China. Therefore, synergetic promotion of inter-city carbon emission control will be the key work of the coordinated development of Beijing, Tianjin, and Hebei in the future [51,52]. It is essential to analyze the synergetic emission reduction effect in the urban agglomeration of Beijing–Tianjin–Hebei on the basis of clarifying the characteristics of the spatial network of carbon emissions in Beijing, Tianjin, and Hebei.

4.2. Analysis of the Synergetic Effect of Carbon Emission Reduction

Based on the synergetic effect model of Formula (8) mentioned above, choosing Beijing as the learning benchmark, the synergetic effect of carbon emission reduction in the urban agglomeration of Beijing–Tianjin–Hebei was calculated from 2007 to 2016. The specific results are shown in Table 2.

Table 2. The synergetic effect of carbon emission reduction in the urban agglomeration of Beijing–Tianjin–Hebei.

City	2007	2008	2009	2010	2011	2012	2013	2014	2015	2016	Mean
Tianjin	−2.404	−2.071	−1.100	0.705	0.827	0.711	−0.457	0.047	−0.424	−0.550	−0.466
Shijiazhuang	−2.562	−1.728	−0.277	0.033	0.862	0.619	−1.025	0.135	−0.287	−0.384	−0.461
Chengde	−2.682	−1.031	−0.922	−0.326	0.995	0.998	0.514	−0.996	−2.040	−3.021	−0.869
Zhangjiakou	−3.378	−1.581	−0.709	0.367	0.969	0.723	−0.471	−0.041	−0.385	−1.859	−0.637
Qinhuangdao	−2.722	−1.969	−0.801	0.262	0.856	0.795	0.663	0.040	−0.508	−3.987	−0.737
Tangshan	−3.116	−1.107	−0.246	0.515	0.613	0.440	−0.483	0.330	0.598	−0.569	−0.302
Langfang	−2.510	−1.662	−0.799	−0.004	0.994	0.844	−0.052	0.107	−2.961	−1.009	−0.705
Baoding	−2.136	−2.123	−0.417	0.418	0.821	0.416	−0.820	0.530	−0.378	−0.199	−0.389
Cangzhou	−2.585	−2.063	−0.669	0.519	0.621	0.654	−0.485	0.597	−0.012	−1.454	−0.488
Hengshui	−1.972	−2.303	−1.037	0.300	0.799	0.542	−0.466	0.619	−0.223	−2.079	−0.580
Xingtai	−0.803	−2.985	−0.753	0.165	0.532	0.464	−0.184	0.876	0.093	−2.126	−0.472
Handan	−2.513	−1.614	−0.697	0.365	0.706	0.380	−0.230	0.626	0.523	−2.421	−0.488
Mean	−2.449	−1.849	−0.702	0.277	0.800	0.632	−0.290	0.239	−0.500	−1.653	−0.550

Based on the results of Table 2, the closer the value is to 1, the higher the synergy between the related city and the central city, Beijing. We can see that the synergetic effect of carbon emission reduction in the urban agglomeration of Beijing–Tianjin–Hebei shows a trend of rising first and then declining. The average level of synergetic effect increased from −2.449 in 2007 to 0.800 in 2011, and then decreased to −1.653 in 2016. This is mainly because Beijing, as the national political, cultural, and technological innovation center, has been capable of controlling its carbon emissions within a reasonable level for a long time, while other cities in the urban agglomeration of Beijing, Tianjin, and Hebei are in the process of industrialization and urbanization and the level of carbon emissions in all other regions shows a trend of continuous growth [9]. With the improvement of economic ties, the interests of related cities and Beijing are more closely linked. If a city continues to follow the original development path of high pollution and high emission, Beijing, which has a larger economic volume, can respond by reducing the purchase of polluting products, thus forcing the city to follow the benchmark of Beijing in terms of their production mode. In the past ten years, the average synergetic effect in the urban agglomeration of Beijing–Tianjin–Hebei was −0.550. There is still much room to improve the overall synergetic level. From the perspective of each city, the synergetic effect of carbon emission reduction is significant. The synergetic effect of carbon emission reduction between Tianjin, Shijiazhuang, Tangshan, Baoding, Xingtai, and other cities and Beijing is relatively high. With the further promotion of a coordinated development strategy of Beijing, Tianjin, and Hebei, a crucial mission in the future will be to change the economic development mode of high investment, high consumption, and high emission and to establish an effective mechanism of coordinated emission reduction through relieving the non-capital functions of Beijing and upgrading the industrial transfer in Beijing.

As mentioned above, structural data usually determine the performance of the attribute data. Based on a clear description of the spatial network of carbon emissions in the urban agglomeration of Beijing–Tianjin–Hebei and an in-depth analysis of the synergetic emission reduction effect in the urban agglomeration of Beijing, Tianjin, and Hebei, this paper chose the annual average of the synergetic emission reduction effect from 2007 to 2016 in the Beijing–Tianjin–Hebei urban agglomeration as the explained variable, respectively, and chose the network intensity, network hierarchy, and network efficiency from 2007 to 2016 as the explanatory variables. We carried out the ordinary least square (OLS) regression, aiming to reveal the impact of the overall structure of the spatial network on the synergetic emission reduction effect. The regression results are shown in Table 3.

Table 3. The regression results of the relationship between the carbon emission network and synergetic emission reduction effect in the urban agglomeration of Beijing–Tianjin–Hebei.

Explanatory Variables	Model (1)	Model (2)	Model (3)
Constant Term	−8.457 **	6.646 *	7.540 **
Network Density	31.809 **	—	—
Network Hierarchy	—	−10.283 *	—
Network Efficiency	—	—	−13.613 **
R^2	0.441	0.173	0.448
F-statistic	6.312	1.671	6.480

Note: ** and * indicate that the effect is significant within the levels of 5% and 10% respectively.

According to the results of Table 3, the regression coefficients of network density, network hierarchy, and network efficiency of the spatial correlation of carbon emissions in the urban agglomeration of Beijing–Tianjin–Hebei are 31.809, −10.283, and −13.613, respectively, which indicate that the spatial network of carbon emissions has a significant impact on the synergetic effect of carbon emission reduction. The increase in network density and the decrease in network hierarchy and network efficiency will significantly strengthen the synergetic effect of carbon emission reduction. This is mainly because the increase in network density means that the number of relationships in the whole network will increase, which will help to limit and narrow the spatial differences of carbon emissions among cities in the network. The decline of the network hierarchy will gradually break down the strict hierarchical structure among cities in the network, and the former subordinate and marginal cities will have more discourse power. Additionally, the decline in network efficiency shows the increase of links in the spatial network of carbon emissions, which will eliminate the comparative advantages of some cities in the network in terms of carbon emissions and in turn increase the synergetic effect of carbon emission reduction.

5. Conclusions and Policy Implications

5.1. Main Conclusions

Based on the data of 13 cities in the urban agglomeration of Beijing–Tianjin–Hebei from 2007 to 2014, this paper adopted the improved gravitational model to characterize the spatial correlation of carbon emissions in the urban agglomeration of Beijing–Tianjin–Hebei. On this basis, the spatial characteristics of the carbon emission network in the urban agglomeration of Beijing–Tianjin–Hebei were investigated by using the analysis method of the social network. Then, a synergetic effect model of carbon emission reduction was constructed to analyze the synergetic emission reduction in the urban agglomeration of Beijing–Tianjin–Hebei. The main conclusions are as follows:

(1) During the research, the spatial correlation network of carbon emissions in the urban agglomeration of Beijing–Tianjin–Hebei presented a typical "center–periphery" structure. From the overall structure of the network, the network density showed a trend of first rising and then declining and the spatial relationship of carbon emissions in the urban agglomeration of Beijing–Tianjin–Hebei still needs strengthening. The results of the network correlation degree show that the network accessibility in the urban agglomeration of Beijing–Tianjin–Hebei is in good condition; the results of the network hierarchy show that in the spatial network of carbon emissions, the hierarchical structure among the cities of Beijing, Tianjin, and Hebei is stricter. Network efficiency showed an oscillating trend, and its value was generally higher. These results indicate that the spatial correlation of carbon emissions has broken through the traditional geographical limitation, showing a complex multi-threaded spatial network correlation, with many overlapping phenomena in the spatial spillover of carbon emissions. The spatial correlation network structure of carbon emissions in the urban agglomeration has gradually stabilized.

(2) Based on the characteristics of network individual centrality, only three of the 13 cities, Tianjin, Beijing, and Baoding, in the urban agglomeration of Beijing–Tianjin–Hebei belong to the recipient

subject in the spatial network correlation of carbon emissions. Beijing and Tianjin are in the center position of the network in the spatial network of carbon emissions. They also play an important role of "intermediary" and "bridge", while other cities are weaker in the spatial network of carbon emissions.

(3) During the research, the synergetic effect of carbon emission reduction in the urban agglomeration of Beijing–Tianjin–Hebei showed a trend of rising first and then declining. The average value of the synergetic effect of carbon emission reduction in the urban agglomeration of Beijing–Tianjin–Hebei was −0.550 over the past ten years. There is still much room to improve the overall synergetic level. From the perspective of each city, the differences in terms of the synergetic effect of carbon emission reduction were quite obvious. The synergetic effect of carbon emission reduction between Tianjin, Shijiazhuang, Tangshan, Baoding, Xingtai, etc., and Beijing was relatively high.

(4) The spatial network of carbon emissions has a significant impact on the synergetic effect of carbon emission reduction. The increase in network density and the decrease in network hierarchy and network efficiency will greatly enhance the synergetic effect of carbon emission reduction. Therefore, enhancing the network structure of spatial correlation of carbon emissions in the urban agglomeration of Beijing–Tianjin–Hebei is an important driving mechanism to narrow the spatial differences of carbon emissions, improve the spatial equity of carbon emissions, and promote the coordinated emission reduction in these regions.

This paper has made a comprehensive analysis and research on the spatial network characteristics and synergetic abatement effect of the carbon emissions in Beijing–Tianjin–Hebei Urban Agglomeration. While we have obtained some research results, and they are of a certain theoretical and practical significance, due to the limitations of personal research ability, there may be still some deficiencies in the paper. Firstly, considering the availability and computability of the data, we only selected the primary energy consumption to calculate the carbon emissions, in the future, we will consider adding more energy consumption data, such as the electricity energy into the formula to improve the carbon emissions calculation. On the other hand, the confirmation of relationships is the key to the analysis of social networks. Currently, this paper uses the improved gravity model to construct the spatial correlation network of carbon emissions in the urban agglomeration of Beijing–Tianjin–Hebei. Whether there will be more advanced methods to determine the network relationships is worth exploring.

5.2. Policy Implications

With the further development of major regional development strategies such as the coordinated development of Beijing, Tianjin, and Hebei, as well as the construction of the Yangtze River Economic Zone, it is an inevitable choice to establish a coordinated mechanism for carbon emission reduction across these regions for the realization of national low-carbon development goals in the future. Comprehensive analysis of the spatial characteristics of carbon emissions and the synergetic emission reduction effects in the urban agglomeration of Beijing–Tianjin–Hebei can provide solutions for the cities in the Beijing, Tianjin, and Hebei provinces to formulate emission reduction policies based on their piratical conditions in the process of synergetic development, in order to achieve the goal of regional synergetic carbon emission reduction [7,53].

Through the research conclusions mentioned above, we can extract the following inspiration for making policies: First, according to the structural characteristics of the "center–periphery" structure of the spatial network of carbon emissions in the urban agglomeration of Beijing–Tianjin–Hebei, we can establish a "lead–follow" type of synergetic emission reduction mechanism. In this type of mechanism, Beijing and Tianjin are the overall leading areas of carbon emission reduction in the urban agglomeration of Beijing–Tianjin–Hebei, while other cities act as the following areas of carbon emission reduction to promote the synergetic control on carbon emission reduction [54]. Secondly, the above research shows that it is difficult to achieve the long-term mechanism of carbon emission reduction by only improving the carbon emission reduction targets of each individual city. It is an inevitable choice to establish a transregional carbon emission reduction coordination mechanism to achieve future carbon emission reduction goals. Thus, we should continuously adjust and optimize the spatial network structure of carbon emissions. At the same time, we should strictly control the total amount of carbon dioxide emissions from the perspective of spatial relations and attach more importance to improving the efficiency of the spatial allocation of resources by using market mechanisms, further boosting the trial establishment of the trading of carbon-emission rights in Beijing and Tianjin, and taking the cities in Hebei province into the scope of trading [1,3]. Finally, we should accelerate the coordinated development strategy of Beijing, Tianjin, and Hebei, establishing the regional sharing mechanism of emission reduction responsibility and the system of emission reduction compensation as soon as possible [55]. While ensuring the economic growth of underdeveloped cities, we should promote the convergence of the level of carbon emissions with the central city, Beijing, enhancing the synergetic effect of carbon emission reduction.

Author Contributions: All authors contributed equally to this work. Specifically, X.L. developed the original idea for the study and designed the methodology. D.F. completed the survey and drafted the manuscript, which was revised by J.L. and Z.Z. All authors read and approved the final manuscript.

Funding: This research was funded by: The National Social Science Foundation of China, Research on the Generating Mechanism Evaluation Method and Governance Model of the Political Cost of Environmental Pollution Problem, No. 16BGL138; the Key Projects of Philosophy and Social Science in Ministry of Education, Study on Some Major Issues in the Implementation of the Beijing–Tianjin–Hebei Synergetic Development Strategy, No. 15JZD021.

Conflicts of Interest: The authors declare no conflict of interest.

References

1. Li, W.; Wang, W.; Wang, Y.; Ali, M. Historical growth in total factor carbon productivity of the Chinese industry—A comprehensive analysis. *J. Clean. Prod.* **2018**, *170*, 471–485. [CrossRef]

2. People.cn. Greenhouse Gas Bulletin: Global Carbon Dioxide Concentrations Hit a New High in 2016. Available online: http://world.people.com.cn/n1/2017/1031/c1002-29618565.html (accessed on 15 December 2018).

3. Tan, X.; Dong, L.; Chen, D.; Gu, B.; Zeng, Y. China's regional CO_2 emissions reduction potential: A study of Chongqing city. *Appl. Energy* **2016**, *162*, 1345–1354. [CrossRef]

4. IEA. CO_2 *Emissions from Fuel Combustion 2013*; IEA: Paris, France, 2014.

5. Chen, J.D.; Cheng, S.L.; Nikic, V.; Song, M.L. Quo Vadis? Major players in global coal consumption and emissions reduction. *Transform. Bus. Econ.* **2018**, *17*, 112–132.

6. Zhao, Q.R.; Chen, Q.H.; Xiao, Y.T.; Tian, G.Q.; Chu, X.L.; Liu, Q.M. Saving forests through development? Fuelwood consumption and the energy-ladder hypothesis in rural Southern China. *Transform. Bus. Econ.* **2017**, *16*, 199–219.

7. Zhang, C.; Lin, Y. Panel estimation for urbanization energy consumption and CO_2 emissions: A regional analysis in China. *Energy Policy* **2012**, *49*, 488–498. [CrossRef]

8. Zhuang, M.; Sheng, J.C.; Webber, M.; Baležentis, T.; Geng, Y.; Zhou, W. Measuring water use performance in the cities along China's South-North water transfer project. *Appl. Geogr.* **2018**, *98*, 184–200.

9. Zhu, Z.; Liu, Y.; Tian, X.; Wang, Y.; Zhang, Y. CO_2 emissions from the industrialization and urbanization processes in the manufacturing center Tianjin in China. *J. Clean. Prod.* **2017**, *168*, 867–875. [CrossRef]

10. Yamaji, K.; Matsuhashi, R.; Nagata, Y.; Kaya, Y. A study on economic measures for CO_2 reduction in Japan. *Energy Policy* **1993**, *21*, 123–132. [CrossRef]

11. Ang, B.W. Is the energy intensity being a less useful indicator than the carbon factor in the study of climate change? *Energy Policy* **1999**, *27*, 943–946. [CrossRef]

12. Bohm, P.; Larsen, B. Fairness in a tradable permit treaty for carbon emission reductions in Europe and the former Soviet Union. *Environ. Resour. Econ.* **1994**, *4*, 219–239. [CrossRef]

13. Janssen, M.; Rotmans, J. Allocation of fossil CO_2 emission rights quantifying cultural perspectives. *Ecol. Econ.* **1995**, *13*, 65–79. [CrossRef]

14. Zhao, G.M.; Zhao, G.Q.; Chen, L.Z.; Sun, H.P. Research on spatial and temporal evolution of carbon emission intensity and its transition mechanism. *China Popul. Resour. Environ.* **2017**, *27*, 84–93. (In Chinese)

15. Jiang, J.J.; Ye, B.; Xie, D.J.; Tang, J. Provincial-level carbon emission drivers and emission reduction strategies in China: Combining multi-layer LMDI decomposition with hierarchical clustering. *J. Clean. Prod.* **2017**, *169*, 178–190. [CrossRef]

16. Jorgenson, A.; Schor, J. Income Inequality and Carbon Emissions in the United States: A State-level Analysis, 1997–2012. *Ecol. Econ.* **2017**, *134*, 40–48. [CrossRef]

17. Zhao, R.; Min, N.; Gen, Y. Allocation of carbon emissions among industries/sectors: Emissions intensity reduction constrained approach. *J. Clean. Prod.* **2017**, *142*, 3083–3094. [CrossRef]

18. Grunewald, N.; Jakob, M.; Mouratiadou, I. Decomposing inequality in CO_2 emissions: The role of primary energy carriers and economic sectors. *Ecol. Econ.* **2014**, *100*, 183–194. [CrossRef]

19. Marbuah, G.; Amuakwa-Mensah, F. Spatial analysis of emissions in Sweden. *Energy Econ.* **2017**, *68*, 383–394. [CrossRef]

20. Wu, H. Analysis on Chinese Provincial Carbon Emission Reduction: Spatial-temporal Patterns, Evolution Mechanisms and Policy Recommendations: Based on the Theory and Method of Spatial Econometrics. *Manag. World* **2015**, *11*, 3–10. (In Chinese)

21. Wang, B.; Sun, Y.F.; Wang, Z.H. Agglomeration effect of CO_2 emissions and emissions reduction effect of technology: A spatial econometric perspective based on China's province-level data. *J. Clean. Prod.* **2018**, *204*, 96–106. [CrossRef]

22. Yan, D.; Lei, Y.L.; Li, L.; Song, W. Carbon emission efficiency and spatial clustering analyses in China's thermal power industry: Evidence from the provincial level. *J. Clean. Prod.* **2017**, *156*, 518–527. [CrossRef]

23. You, W.H.; Lv, Z.K. Spillover effects of economic globalization on CO_2 emissions: A spatial panel approach. *Energy Econ.* **2018**, *73*, 248–257. [CrossRef]

24. Sun, H.; Liu, Y. Analysis on inter-regional differences of carbon emissions loss and profit deviation in China. *Manag. Rrev.* **2016**, *10*, 89–96. (In Chinese)

25. Wang, Y.; Zhao, T.; Wang, J.; Guo, F.; Kan, X.; Yuan, R. Spatial analysis on carbon emission abatement capacity at provincial level in China from 1997 to 2014: An empirical study based on SDM model. *Atmos. Pollut. Res.* **2019**, *10*, 97–104. [CrossRef]

26. Zhang, Y.; Wua, Q.; Zhao, X.; Hao, Y.; Liu, R.; Yang, Z.; Lu, Z. Study of carbon metabolic processes and their spatial distribution in the Beijing-Tianjin-Hebei urban agglomeration. *Sci. Total Environ.* **2018**, *7*, 1630–1642. [CrossRef] [PubMed]

27. Wang, F.; Gao, M.N.; Liu, J.; Fan, W.N. The spatial network structure of China's regional carbon emissions and its network effect. *Energies* **2018**, *11*, 2076. [CrossRef]

28. Sun, Y.; Liu, H.; Liu, C. Research on spatial association of provinces carbon emissions and its effects in China. *Shanghai Econ. Res.* **2016**, *2*, 82–92. (In Chinese)

29. Yang, G.Y.; Wu, Q.; Xu, Y. Researches of China's regional carbon emission spatial correlation and its determinants: Based on the method of social network analysis. *J. Bus. Econ.* **2016**, *4*, 56–78. (In Chinese)

30. Chen, J.; Zhao, A.; Zhao, Q.; Song, M.; Baležentis, T.; Streimikiene, D. Estimation and factor decomposition of carbon emissions in China's tourism sector. *Probl. Ekorozw.* **2018**, *13*, 91–101.

31. Li, J.C.; Xiang, Y.W.; Jia, H.Y.; Chen, L. Analysis of total factor energy efficiency and its influencing factors on key energy-intensive industries in the Beijing-Tianjin-Hebei region. *Sustainability* **2018**, *10*, 111. [CrossRef]

32. Yang, Q.; Wang, X.Z.; Ma, H.M. Assessing green development efficiency of municipalities and provinces in China integrating models of super-efficiency DEA and Malmquist index. *Sustainability* **2015**, *7*, 4492–4510. [CrossRef]

33. Li, J.; Chen, S.; Wan, G.H.; Fu, C.M. Study on the spatial correlation and explanation of regional economic growth in China-Based on analytic network process. *Econ. Res.* **2014**, *11*, 4–16. (In Chinese)

34. Schröter, B.; Hauck, J.; Hackenberg, I.; Matzdorf, B. Bringing transparency into the process: Social network analysis as a tool to support the participatory design and implementation process of Payments for Ecosystem Services. *Ecosyst. Serv.* **2018**, *34 Pt B*, 206–217. [CrossRef]

35. Rödder, W.; Brenner, D.; Kulmann, F. Entropy based evaluation of net structures—Deployed in Social Network Analysis. *Expert Syst Appl.* **2014**, *41*, 7968–7979. [CrossRef]

36. Zhang, G.L.; Zhang, N.; Liao, W.M. How do population and land urbanization affect CO_2 emissions under gravity center change? A spatial econometric analysis. *J. Clean. Prod.* **2018**, *202*, 510–523. [CrossRef]

37. Maharani, W.; Gozali, A.A. Collaborative Social Network Analysis and Content-based Approach to Improve the Marketing Strategy of SMEs in Indonesia. *Procedia Comput. Sci.* **2015**, *59*, 373–381. [CrossRef]

38. Rios, V.; Gianmoena, L. Convergence in CO_2 emissions: A spatial economic analysis with cross-country interactions. *Energy Econ.* **2018**, *75*, 222–238. [CrossRef]

39. Balado-Naves, R.; Baños-Pino, J.F.; Mayor, M. Do countries influence neighbouring pollution? A spatial analysis of the EKC for CO_2 emissions. *Energy Policy* **2018**, *123*, 266–279. [CrossRef]

40. Badi, S.; Wang, L.S.; Pryke, S. Relationship marketing in Guanxi networks: A social network analysis study of Chinese construction small and medium-sized enterprises. *Ind. Mark. Manag.* **2017**, *60*, 204–218. [CrossRef]

41. Brooks, B.; Hogan, B.; Ellison, N.; Lampe, C.; Vitak, J. Assessing structural correlates to social capital in Facebook ego networks. *Soc. Networks* **2014**, *38*, 1–15. [CrossRef]

42. Zhang, L.L.; Xiong, L.C.; Cheng, B.D.; Yu, C. How does foreign trade influence China's carbon productivity? Based on panel spatial lag model analysis. *Struct. Chang. Econ. Dyn.* **2018**, *47*, 171–179. [CrossRef]

43. Warner, J.; Zawahri, N. Hegemony and Asymmetry: Multiple-chessboard Games on Transboundary Rivers. *Int. Environ. Agreem.* **2012**, *12*, 215–229. [CrossRef]

44. Zhang, L.J.; Rong, P.J.; Qin, Y.C.; Ji, Y.Y. Does Industrial Agglomeration Mitigate Fossil CO_2 Emissions? An Empirical Study with Spatial Panel Regression model. *Energy Procedia* **2018**, *152*, 731–737. [CrossRef]

45. Li, W.C.; Yan, Y.H.; Tian, L.X. Spatial Spillover Effects of Industrial Carbon Emissions in China. *Energy Procedia* **2018**, *152*, 679–684.

46. Cerqueira, P.; Martins, R. Measuring the Determinants of Business Cycle Synchronization Using a Panel Approach. *Econ. Lett.* **2009**, *102*, 106–108. [CrossRef]

47. Can, M.; Gozgor, G. The impact of economic complexity on carbon emissions: Evidence from France. *Environ. Sci. Pollut. Res.* **2017**, *24*, 16364–16370. [CrossRef] [PubMed]

48. Han, Y.M.; Long, C.; Geng, Z.Q. Carbon emission analysis and evaluation of industrial departments in China: An improved environmental DEA cross model based on information entropy. *J. Environ. Manag.* **2018**, *205*, 298–307. [CrossRef] [PubMed]

49. Ikeda, K.; Murota, K.; Takayama, Y. Stable economic agglomeration patterns in two dimensions: Beyond the scope of central place theory. *J. Reg. Sci.* **2017**, *57*, 132–172. [CrossRef]

50. Neves, S.A.; Marques, A.C.; Fuinhas, J.A. Is energy consumption in the trans-port sector hampering both economic growth and the reduction of CO_2 emissions? A disaggregated energy consumption analysis. *Transp. Policy* **2017**, *59*, 64–70. [CrossRef]

51. Riti, J.S.; Song, D.Y.; Shu, Y. Decoupling CO_2 emission and economic growth in China: Is there consistency in estimation results in analyzing environmental Kuznets curve? *J. Clean. Prod.* **2017**, *166*, 1448–1461. [CrossRef]

52. Zhang, Z.; Craft, C.B.; Xue, Z.; Tong, S.; Lu, X. Regulating effects of climate, net primary productivity, and nitrogen on carbon sequestration rates in temperate wet-lands, Northeast China. *Ecol. Indic.* **2016**, *70*, 114–124. [CrossRef]

53. Tao, Y.; Li, F.; Wang, R.; Zhao, D. Effects of land use and cover change on terrestrial carbon stocks in urbanized areas: A study from Changzhou, China. *J. Clean. Prod.* **2015**, *103*, 651–657. [CrossRef]

54. Criado, C.O.; Grether, J.M. Convergence in per capita CO_2 emissions: A robust distributional approach. *Resour. Energy Econ.* **2011**, *33*, 637–665. [CrossRef]

55. Aichele, R.; Felbermayr, G. Kyoto and carbon leakage: An empirical analysis of the carbon content of bilateral trade. *Rev. Econ. Stat.* **2015**, *97*, 104–115. [CrossRef]

 *sustainability*

Article

Standardization of the Evaluation Index System for Low-Carbon Cities in China: A Case Study of Xiamen

Longyu Shi [1], Xueqin Xiang [1,2], Wei Zhu [1,2] and Lijie Gao [1,*]

[1] Key Laboratory of Urban Environment and Health, Institute of Urban Environment,
     Chinese Academy of Sciences, 1799 Jimei Road, Xiamen 361021, Fujian, China; lyshi@iue.ac.cn (L.S.);
     xqxiang@iue.ac.cn (X.X.); wzhu@iue.ac.cn (W.Z.)
[2] University of Chinese Academy of Sciences, Beijing 100049, China
[*] Correspondence: ljgao@iue.ac.cn; Tel.: +86-61-90-676

Received: 29 August 2018; Accepted: 12 October 2018; Published: 18 October 2018

Abstract: The construction of a reasonable evaluation index system for low-carbon cities is an important part of China's green development strategy in urban areas. In this study, based on the theoretical framework for the concept of low-carbon cities, the perspectives from three index systems—that is, the Drivers, Pressures, State, Impact, Response model of intervention (DPSIR), a complex ecosystem, and a carbon source/sink process—were integrated to extract common indicators from existing evaluation index systems for low-carbon cities. Subsequently, a standardized evaluation index system for low-carbon cities that contained five indicators—carbon emission, low carbon production, low carbon consumption, low-carbon policy, and social economic development—was established. Thereafter, Xiamen was selected for an empirical analysis by determining the indicator weight with an entropy weight method and by carrying out a comprehensive evaluation using a linear summation model. The results showed that the weights of the five selected primary indicators for the evaluation of low-carbon cities were: low-carbon production > low-carbon consumption > social economic development > carbon emission > low-carbon policy. Among the secondary indicators, the average entropy weight of "pollution emission" was the highest at 0.1591, while the average entropy weight of "urbanization rate" was the lowest at 0.0360. Furthermore, the comprehensive index of low-carbon development in 2015 was higher than that in 2010, while the rate of economic growth was greater than the growth rate of carbon emission, which indicated that the relative decoupling of economic growth from carbon emission was basically achieved.

Keywords: low-carbon city; evaluation index; standardization; entropy weight method; level

1. Introduction

In recent years, global warming has led to the gradual adoption of developmental models involving green, low-carbon, and circular economies [1]. China has become the largest energy producer and consumer in the world [2]. In 2016, China's carbon emissions accounted for 27.3% of the world's total carbon emissions, making China the world's largest carbon emitter [3]. Cities are the main contributors to carbon emissions, accounting for 75% of total emissions, and this proportion continues to rise with urbanization [4]. The UN's publication, "Transforming our World: the 2030 Agenda for Sustainable Development", contains 17 sustainable development goals, and one of the key objectives is to take urgent action to address climate change and its impacts and promote sustainable development [5]. In the 2015 United Nations Climate Change Conference held in Paris, China submitted the Intended Nationally Determined Contributions (INDC) documents, which stated that China would reach its peak in CO_2 emissions around 2030 and would strive to decrease its CO_2 emission per unit Gross Domestic Product (GDP) in 2030 by 60–65% compared to that in 2005 [6].

In this context, the low-carbon development model has become the best choice for long-term global development [7]. Scholars have conducted extensive research focused on low-carbon development, and the evaluation of low-carbon cities has become a popular area of research. The study reviewed the four representative international low-carbon indicator systems and 14 representative Chinese low-carbon indicator systems. The primary indicators selected by scholars in constructing the index system are mainly related to energy, economy, society, and environment. The scope of evaluation is extensive, including an evaluation of the low-carbon development level of the world, the country, and the city. The evaluation results provide a realistic basis for city decision makers and managers to formulate and manage low-carbon development of cities, and may have a great role in promoting the low-carbon development of cities (Table 1). However, the lack of adequate standards related to low-carbon cities significantly constrains the strategic development, planning, and construction of low-carbon cities.

Table 1. The low-carbon evaluation index system at home and abroad.

Country	Year	Source	Indicator Category	Specific Evaluation Aspect
Saudi Arabia	2018	Azizalrahman, H.; Hasyimi, V. [8]	Economy, energy, land use, carbon and environment, transportation, waste and water	Evaluate the low-carbon development of ten low-carbon pilot cities at home and abroad.
Cambodia	2017	Hak, M.; Matsuoka, Y.; Gomi, K. [9]	Demography, economy, energy, transportation and cross sector	Collect information from these sectors in Cambodia and formulate strategies for low-carbon development.
Malaysia	2015	Tan, S.T.; Yang, J.; Yan, J.Y. [10]	Economic, energy pattern, technology, social and living, carbon and environment, urban accessibility and waste	It is used to evaluate the low-carbon development level of 10 major domestic and foreign cities.
Latvia	2015	Kalnins, S.N.; Blumberga, D.; Gusca, J. [11]	Technological, economic, social, environmental, climate	—
America	2012	Zhou, N.; He, G.; Williams, C.; Fridley, D. [12]	Energy/climate, water, air quality, waste, mobility, economic health, land use and social health	—
Germany	2010	Lu, Q. [13]	Environment, economy, social culture and function, technology and process	Evaluating the overall performance of urban areas.
Global	2009	Pamlin, D. [14]	direct emissions, embedded emissions and global solutions	LCCI (Low-Carbon City Index): Evaluating the low-carbon development of global cities.
China	2018	Zhang, L.P.; Zhou, P. [15]	Economic, living quality, environment, consumer behavior	The low-carbon development level of 40 cities in China is evaluated and it is found that the low-carbon development of coastal cities in China is generally superior to other regions.
	2018	Yang, X.; Wang, X.C.; Zhou, Z.Y. [16]	Carbon emission per capita, carbon emission per unit of Gross Domestic Product (GDP), GDP per capita, population, urbanization rate, proportion of tertiary industry, main functional zone	GDP per Capita and Carbon emission per capita are used to evaluate the low-carbon development of 36 low-carbon pilot cities in the country and divide them into four types: Leading Cities, Developing Cities, Latecomer Cities and Exploring Cities.

Table 1. *Cont.*

Country	Year	Source	Indicator Category	Specific Evaluation Aspect
	2018	Du, H.B.; Chen, Z.N.; Mao, G.Z.; Li, R.Y.M.; Chai, L.H. [7]	Society, economy, energy, environment	Evaluate the low-carbon development of 30 provinces in the country from 2003 to 2013, and divide these cities into the highest, middle, and lowest three levels.
	2017	Ohshita, S.; Zhang, J.; Yang, L.; et al. [17]	CGLCI (China Green Low-Carbon City Index): economy, energy & carbon, environment & land use, policy & outreach	Evaluating the status of green and low-carbon development for a large number of Chinese cities.
	2013	Price, L.; Zhou, N.; Fridley, D.; Ohshita, S.; Lu, H.Y.; Zheng, N.; Fino-Chen, C. [18]	Residential buildings sector, commercial buildings sector, industry sector, transport sector, power sector	Evaluation of low-carbon development of provincial and municipal cities from the perspective of a variety of sectors.
	2010 2011	Fu, Y.; Liu, Y.J.; Wang, Y.L. [19] Hua, J.; Ren, J. [20]; Xin, L. [21]	Economy, society, environment	—
	2017	Zhu, J.; Liu, X.M.; Zhang, Y. [22]	DPSIR model (involving energy consumption, carbon emissions, industry, transportation, construction, waste disposal)	—
	2010	Shao, C.F.; Ju, M.T. [23]		
	2012	Zhang, X.P.; Liu, J.; Fang, T. [24]		
	2011	Zhang, L.; Chen, K.L.; Cao, S.K. [25]	Carbon source (energy, industry, construction, transportation), carbon sink (land, green space)	—
	2011	Chu, C.L.; Ju, M.T.; Wang, Y.N.; Wang, Y.S. [26]		
	2011	The Media Alliance of China Low-Carbon Economy. [27]	Low-carbon development planning media communication, new and renewable sources of energy, low-carbon product application rate, percentage of greenery coverage, low-carbon commuting, low-carbon buildings, air quality, urban direct carbon reduction, public satisfaction and support rate, veto power	Chinese mainland cities.
	2010	Chatham House; Chinese Academy of Social Sciences; Energy Research Institute; Jilin University; E3G. [28]	Low-carbon productivity, low-carbon consumption, low-carbon resources, low-carbon policy	Evaluating and managing the city's low-carbon construction, which is used in Jilin's low-carbon development plan.

Therefore, based on a systematic review of existing index systems for low-carbon cities, this study comprehensively compared the advantages and disadvantages of each index system. In addition, by extracting reasonable common indicators and using relevant standards of urban construction promulgated by the Chinese government as a reference, this study attempted to construct a standardized evaluation index system for low-carbon cities. It is expected that this index system can be used by urban-level management departments to evaluate low-carbon development within cities, as well as by national-level management departments to evaluate and compare low- carbon development among different cities, so that a better path for the development of low-carbon cities can be explored.

2. Construction of the Index System

2.1. Theoretical Basis for Standardization of the Evaluation Index System

We conducted an in-depth comparison and systematic analysis of representative evaluation index systems for low-carbon cities in China and elsewhere. Subsequently, all index systems were classified into three categories according to the perspective from which the index system is constructed, while all indicators in each category of index systems were then summarized according to their frequency and commonality. It was found that the index systems were mainly constructed from the following three perspectives:

2.1.1. Orientation from the Drivers-Pressures-State-Impact-Response (DPSIR) Model of Intervention

Some evaluation index systems for low-carbon cities were constructed based on the DPSIR model, which emphasizes economic operations and their impact on the environment, as well as the relationships between economic operations and the environment. The DPSIR model is comprehensive, systematic, holistic, and flexible. Most of its "Drivers" are selected from the aspects of GDP, industrial output, and disposable income, while its "Pressures" are mainly selected from the aspects of resource consumption, energy consumption, and the mode of consumption. The "State" of the DPSIR model is mainly selected from the aspects of pollutant discharge and industry/energy structure, the "Impact" is mainly selected from the aspects of social development, ecological environment, and public evaluation, and the "Response" is mainly selected from the aspects of pollution control and carbon sink capacity (Table 2).

Table 2. Evaluation indicators of low-carbon cities, established based on the DPSIR model.

Category	Most Frequently Selected Indicator
Drivers	Gross regional product/per capita gross regional product, per capita disposable income of urban residents, urbanization rate
Pressures	Unit GDP energy/electricity/water consumption, per capita energy/water/electricity consumption, energy/electricity/water consumption per unit of industrial added value, public transportation vehicles per 10,000 people
State	Unit GDP carbon emissions, per capita carbon emissions, percentage of tertiary industry in GDP
Impact	Number of days with good or adequate ambient air quality, Engel coefficient, public perception of low-carbon cities
Response	Forest coverage, green coverage of built-up areas, per capita public green space

Source Index: [22–24,29].

The evaluation index systems constructed based on the DPSIR model summarizes the development level of low-carbon cities in terms of Drivers, Pressures, State, Impact, and Response. Such index systems help to facilitate an intuitive understanding of the assuring factors, problems, status quo, and countermeasures required to maintain the development of low-carbon cities. As a result, such index systems can solve the problems faced by decision-makers with regard to developing effective urban management policies, which are difficult to achieve because it is often unclear which departments are responsible for low-carbon development, and the level of low-carbon development in specific sectors can be difficult to determine.

2.1.2. Orientation from the Complex Ecosystem

Based on the concept of the complex ecosystem, the index system for the evaluation of low-carbon cities can be set up by configuring indicators for the three subsystems of a city—that is, the social, economic, and environmental systems.

As a complex ecosystem, the indicators of low-carbon cities mainly characterize low-carbon development in the social, economic, and environmental subsystems. The indicators for the economic system are mainly relevant to total production output and per capita disposable income of urban residents. The indicators for the social system are mainly related to aspects of quality of life, consumption mode, transportation systems, and social security. The indicators for the environmental system are related to aspects of energy consumption structure, resource consumption, pollution emissions, carbon emissions, and ecological environment quality (Table 3).

Table 3. Evaluation indicators of low-carbon cities, established based on the complex ecosystem perspective.

Category	Most Frequently Selected Indicator
Economic system	GDP per capita, per capita disposable income of urban residents
Social system	Urbanization rate, Engel coefficient, number of public transportation vehicles per 10,000 people, proportion of clean energy vehicles, share of travel by public transportation, registered urban unemployment rate
Environmental system	Proportion of non-fossil energy consumption in total primary energy consumption, per capita energy/electricity/water consumption, sulfur dioxide, ammonia nitrogen and nitrogen oxide emissions and chemical oxygen demand, total carbon emissions, per capita carbon emissions, carbon emission intensity, forest coverage, green coverage in built-up areas, per capita public green space

Source index: [19–21,30–34].

It is easier for urban managers to understand the specific conditions of various sectors or industries in cities and to divide responsibilities among relevant decision makers if a low-carbon city is treated as a complex ecosystem, so that the level of low-carbon development can be analyzed systematically based on the social, economic, and environmental subsystems. However, such systems are only applicable to the internal evaluation of urban development, and are not applicable to the external evaluation of urban development. From the perspective of existing index systems, the indicators selected by each system and individual departments are not homogeneous, and the nature of such indicators does not consider all aspects of Drivers, Pressures, State, Impact, and Response.

2.1.3. Orientation from the Carbon Source/Sink Process Perspective

The index system for the evaluation of low-carbon cities can be established from the perspective of the carbon source/sink process (resources, production, consumption, emissions, and treatment).

In essence, the construction of an index system from the perspective of a carbon source/sink is an analysis of the level of low-carbon development in cities, with a focus on resources, production, consumption, emissions, and treatment. The process of production using resources and subsequent consumption and emissions can be considered as carbon source generation, and treatment is the carbon sink process. The carbon sources in cities mainly come from the industrial, transportation, and construction sectors, as well as from the household consumption of residents. In previous research, the indicators to characterize carbon sources have mainly been selected from five aspects: energy, industry, transportation, construction, and household consumption of residents. In contrast, the indicators used to represent carbon sinks have mainly been related to aspects of green space (Table 4).

Table 4. Evaluation indicators of low-carbon cities, established based on the carbon source/sink process.

Category	Most Frequently Selected Indicator
Carbon source	Energy consumption per unit of GDP, proportion of clean energy vehicles, number of public transport vehicles owned by 10,000 people, share of travel via public transportation, proportion of energy-saving buildings, carbon emissions per unit of building area, carbon emissions per capita, per capita disposable income of urban residents, Engel coefficient, per capita water/electricity consumption of residents, and per capita energy consumption
Carbon sink	Per capita green public area, green area coverage of built-up area, forest coverage

Source index: [25,26].

The statuses of resource production and consumption, as well as pollution discharge and treatment, can be more clearly understood if the carbon level in cities is analyzed with a focus on resources, production, consumption, emission, and treatment. Therefore, such an approach is applicable for comparisons of the levels of low-carbon development among different low-carbon cities. However, based on the analysis using the DPSIR model, the index systems constructed from this perspective lack indicators characterizing the Drivers of low-carbon urban development. In addition, the low-carbon development profile of specific urban sectors is not fully reflected in such an approach—that is, indicators applicable to the internal evaluation of cities are not fully reflected.

In summary, the index systems constructed from the three perspectives described above have distinct advantages and disadvantages. Index systems generated from a single perspective can no longer meet the growing demand for more scientific and reasonable evaluation of the low-carbon development of cities. Therefore, we integrated the three perspectives described above to establish an evaluation index system for low-carbon cities that can be used for both internal (i.e., vertical) and external (i.e., horizontal) evaluation.

2.2. Low-Carbon Evaluation Index System Constructed from Three Perspectives

Based on our review of current evaluation index systems, and in strict accordance with pertinent scientific, systematic, hierarchical, dynamic, and operable principles, an index system was established from the perspectives of the DPSIR model, the complex ecosystem, and the carbon source/sink process. This model was constructed in reference to recognized index systems in China and elsewhere. The evaluation of the model was focused on low-carbon development, eco-cities, green cities, and sustainable development. After several rounds of systematic comparison, analysis and screening of existing index systems, ISO37120, the "Low-carbon eco-city evaluation tool for China", the "Plan for low-carbon development in Jilin city", the "Media alliance standard of China's low-carbon economy", the "China Green Development Index", and the "Index system for the evaluation of low-carbon city construction in China" were selected as the primary references from which common indicators corresponding to low-carbon urban construction were extracted [12,22,27,28,35,36]. Subsequently, individual indicators were adjusted by consulting with experts in this field to construct a final standardized index system for low-carbon evaluation, which included five aspects: carbon emissions, low-carbon production, low-carbon consumption, low-carbon policies, and socio-economic development (Figure 1).

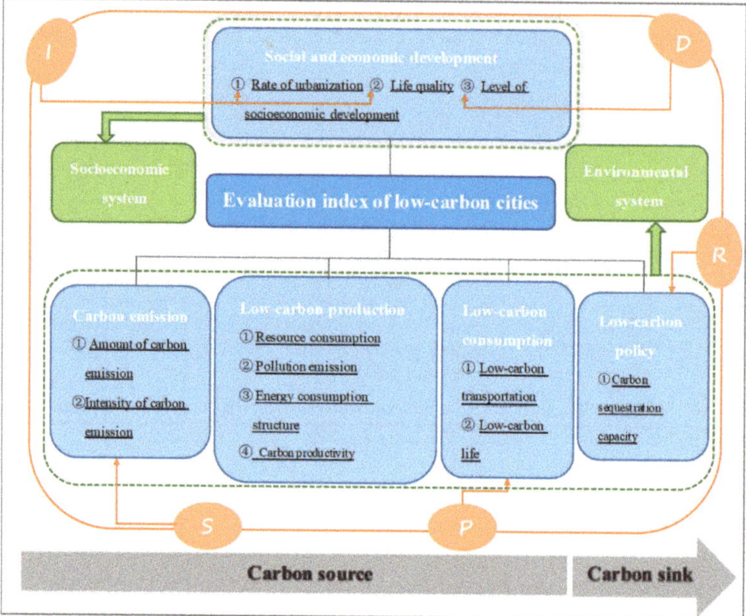

Figure 1. Framework for the evaluation index system of low-carbon cities.

From the range of evaluation indicators, the external evaluation selected carbon emission indicators to evaluate the low-carbon development of different cities, while the internal evaluation selected a set of indicators to reflect the low-carbon construction of different industries and departments within a city. The evaluation conducted with these indicators showed that they reflect the current status of low-carbon development, as well as low-carbon strategies and approaches for further improvement. The attributes of these indicators are also balanced with regard to positive and negative aspects. The positive indicators reflect social and economic development, carbon source and sink capacities, and energy conservation and environmental protection. The negative indicators reflect carbon emissions, pollution emissions, and resource consumption (Table 5).

Table 5. Evaluation index system for low-carbon cities.

Primary Indicator	Secondary Indicator	Tertiary Indicator	Indicator Attribute
Carbon emissions	Amount of carbon emissions	Total carbon emissions	−
		Per capita carbon emissions	−
	Intensity of carbon emissions	Carbon emissions per unit of GDP	−
Low-carbon production	Resource consumption	Energy consumption per unit carbon emissions	−
		Water consumption per unit carbon emissions	−
		Electricity consumption per unit carbon emissions	−
	Pollution emissions	Ammonia nitrogen emissions per unit of carbon emissions	−
		Chemical oxygen demand per unit of carbon emissions	−
		Nitrogen oxide emissions per unit of carbon emissions	−
		Sulfur dioxide emissions per unit of carbon emissions	−
	Energy consumption structure	Proportion of non-fossil energy in primary energy consumption	+
	Carbon productivity	GDP per unit carbon emissions	+

Table 5. *Cont.*

Indicator Name			Indicator Attribute
Primary Indicator	Secondary Indicator	Tertiary Indicator	
Low-carbon consumption	Low carbon transportation	Proportion of clean energy vehicles	+
		Share of travel via public transportation	+
		Public transportation vehicles owned by every 10,000 people	+
	Low carbon life	Per capita household water consumption	−
		Per capita household electricity consumption	−
		Per capita production of household garbage	−
Low-carbon policy	Carbon sink capacity	Per capita green public space	+
		Green coverage in built-up areas	+
		Forest coverage	+
Social and economic development	Urbanization	Urbanization rate	−
	Quality of life	Engel coefficient	−
		Average life expectancy	+
		Registered urban unemployment rate	+
	Level of economic development	Per capita GDP	+
		Annual per capita disposable income of urban residents	+

Note: (+) in the table indicates a positive indicator and (−) indicates a negative indicator.

3. Methodology

The evaluation method is mainly reflected in the two key links of index weight determination and comprehensive evaluation. There are two main methods to determining the weight of indicators: subjective evaluation and objective evaluation. Subjective evaluation mainly focuses on the Analytic Hierarchy Process and the Delphi method. For example, Hua evaluated the low-carbon construction of 13 prefecture-level cities in Jiangsu Province based on an analytic hierarchy process, and the Media Alliance of China's Low-Carbon Economy used the Delphi method to evaluate the low-carbon development of mainland Chinese cities [20,27]. Objective evaluation methods mainly use the correlation coefficient method, the entropy weight method, and the factor analysis method. Tan used the entropy weight method to evaluate the low-carbon development of 10 domestic and foreign cities, and Yi used the factor analysis method to make an empirical analysis of the development level of low-carbon cities in the six central provincial capitals in 2008 [10,37]. In general, the subjective evaluation method determines the importance of each index according to subjective judgment, which has the advantages of a clear concept, simplicity, and feasibility; however, it is more easily interfered with by subjective factors. According to the standardized data of each index, the objective evaluation method automatically entrust weight according to certain rules. Although its advantages are rigorous calculation and objective evaluation, the results can change along with the data and the stability is poor. Furthermore, the comprehensive evaluation model mainly concentrates on Technology for Order Preference by Similarity to an Ideal Solution (TOPSIS), the Analytic Hierarchy Process (AHP), Fuzzy Comprehensive Evaluation (FCE), the Synthetical Index method, and the Content Analysis method, where comparisons of each method is presented in Table 6. The comprehensive evaluation model and weight-quantifying methods have their own advantages and disadvantages. In the actual evaluation, it is necessary to select the appropriate evaluation method according to the actual conditions.

Table 6. Comparison of various low-carbon evaluation models.

Method	Method Description	Advantages	Limitations	Reference
TOPSIS	According to the approaching degree of limited evaluation objects and idealized goals, the relative merits of existing objects are evaluated.	Making full use of the original information, the results are likely to be consistent with reality; is able to sort out the merits and faults of each evaluation object.	Low sensitivity	Zhang, X.L. [38]
AHP	This refers to the decision-making method which decomposes the elements related to the decision in regard to goals, criteria, schemes, and so on, and facilitates qualitative and quantitative analysis on this basis.	Systematic; results are simple and clear; less quantitative data needed	There are less quantitative data, more qualitative components, and is less convincing.	Yang, Y.F. [39]
FCE	According to the membership degree theory of fuzzy mathematics, qualitative evaluation is transformed into quantitative evaluation—that is, fuzzy mathematics makes an overall evaluation of things or objects restricted by various factors.	Good for solving fuzzy and difficult problems; is also suitable for solving various non-deterministic problems; strong systematicness.	Evaluation subjectivity is obvious, and the evaluation results are not comprehensive enough	Ma, L.; Liu, X.G.; Liu, Z.W. [40]
Synthetical Index	This is done through the comparison of two comprehensive total amounts to comprehensively reflect the total change degree of multiple individuals.	The evaluation process is systematic and comprehensive, the calculation is simple, and the data can be fully utilized.	The original data needs to be complete and the reliance on comparative standards is too strong; evaluation result lacks certain intuition	Wang, Y.Z.; Zhou, Y.Y.; Deng, X.Y. [41]
Content Analysis	Non-quantitative literature materials are converted into quantitative data, the content of the literature is quantitatively analyzed, and judgement and inference about the facts are made according to these data.	Systematic, objective, and quantitative	The classification and the operation is complicated	Zhou, G.H.; Singh, J.; Wu, J.C.; Sinha, R.; Laurenti, R.; Frostell, B. [29]

3.1. Carbon Emission Calculated Based on IPCC Assessment

Based on the IPCC Guidelines for Greenhouse Gas Inventories, calculation of urban carbon emissions primarily involves greenhouse gases generated by activities in five areas: energy activities, industrial processes, changes in land use, forestry and agricultural activities, and waste disposal [42]. The calculation principle is:

$$\text{Emissions} = \text{Activity level} \times \text{Emission factor}$$

The data on activity level were directly obtained or calculated from the statistical yearbook. The emission factor data were mainly obtained according to the "Guidelines for the Accounting Tools of Urban Greenhouse Gas (Test Version 1.0)" and relevant studies [43,44].

3.2. An Entropy Weight Method to Determine the Weight of Different Indicators

The entropy weight method is a mathematical method for calculating a comprehensive index based on the comprehensive consideration of the amount of information provided by various factors [45,46]. As an objective and comprehensive weighting method, the entropy weight method mainly determines the weight according to the amount of information transmitted by each indicator to the decision maker. For a certain indicator X_j, a larger difference in the values of X_{ij} indicates a greater role of this indicator in the comprehensive evaluation. If the values of an indicator are all equal, this indicator does not play a role in the comprehensive evaluation. The specific steps are as follows:

(1) Establish an original matrix, X

$$X = \begin{bmatrix} X_{11} & X_{12} \ldots & X_{1n} \\ X_{21} & X_{22} \ldots & X_{2n} \\ \ldots & \ldots \ldots \ldots \\ X_{m1} & X_{m2} & \ldots & X_{mn} \end{bmatrix}_{m \times n} \tag{1}$$

where m is the number of objects to be evaluated, n is the number of evaluation indicators, and X_{ij} is the evaluation value of object i under indicator j, $i = 1, \ldots, m$, $j = 1, \ldots, n$, $m = 2$, and $n = 27$.

(2) Matrix the indicators according to the same ratio and calculate the weight p_{ij} of indicator j in protocol i:

$$p_{ij} = \frac{X_{ij}}{\sum_{i=1}^{m} X_{ij}} \tag{2}$$

(3) Calculate the entropy value e_j of indicator j:

$$e_j = \frac{-1}{\ln(m)} \sum_{i=1}^{m} (p_{ij} \times \ln p_{ij}) \tag{3}$$

Note: If $p_{ij} = 0$, define $p_{ij} \times \ln p_{ij} = 0$.

(4) Determine the entropy weight w_j of indicator j:

$$w_j = \frac{1 - e_j}{\sum_{j=1}^{n} (1 - e_j)} \tag{4}$$

3.3. Comprehensive Evaluation Using a Linear Summation Model

The so-called linear summation model involves multiplying the weights of indicators by the processed data, followed by a simple summation [38]. It is performed as follows:

$$Y_k = \sum_{j=1}^{n} w_j X_j \tag{5}$$

where Y_k is the comprehensive evaluation level of object k, w_j is the weight of object k under indicator j, X_j is the normalized value of object k under indicator j, and n is the number of indicators for the evaluation object.

3.4. Overview of the Study Area

Xiamen is located in the central part of the west bank of the Taiwan Strait, at the center of the Golden Triangle of southern Fujian (north latitude 24°23′ to 24°54′ and east longitude 117°52′ to 118°26′). Xiamen has a tropical monsoon climate with long, hot, humid summers and warm winters. The annual level of precipitation in Xiamen is relatively high, and it receives a relatively large amount of solar radiation. At the end of 2015, Xiamen had a total population of 3.86 million, including an urban population of 1.68 million. In 2010, the National Development and Reform Commission of China issued the "Notice on Launching Pilot Projects for Low-Carbon Provinces and Low-Carbon Cities", which included Xiamen as one of the "five provinces and eight cities" in the pilot study. Therefore, the analysis of Xiamen's low-carbon development between 2010 and 2015 has practical guiding significance for the construction of low-carbon cities in Xiamen and across China.

3.5. Data Sources

The data used in this study were either obtained directly from statistical yearbooks, literature, and the websites of relevant departments, or by calculation. The yearbooks used in this study included the "Xiamen Special Economic Zone Yearbook 2011" and "Xiamen Special Economic Zone Yearbook 2016", the "Fujian Statistical Yearbook 2011" and "Fujian Statistical Yearbook 2016", and the "Fujian Energy Balance Sheet 2011" and "Fujian Energy Balance Sheet 2016".

4. Results

4.1. Entropy Weight of Indicators and Changes in Entropy Weight

By selecting the profile of low-carbon development in Xiamen between 2010 and 2015 as the research object, five primary indicators, 12 secondary indicators, and 27 tertiary indicators were selected, and the weights of secondary indicators in 2010 and 2015 were calculated by the entropy weight method (Table 7). From the overall situation of indicator weights, the weights of indicators for pollution emissions, low-carbon life, and resource consumption were relatively higher than those of other indicators. From the perspective of changes in indicator weights, the weights of indicators for carbon emissions and pollution emissions decreased over time, while the weights of indicators for resource consumption, low-carbon life, and low-carbon transportation increased significantly over time. Among these indicators, the negative indicator "pollution emission" had the highest average entropy weight (0.1591). Therefore, this indicator played the most important role in the comprehensive evaluation of low-carbon development in Xiamen. The indicator "urbanization rate" had the smallest average entropy weight (0.0360).

Table 7. Entropy weight of evaluation indicators.

Year	2010	2015	Average
Carbon Emissions	0.0733	0.0714	0.0722
Intensity of Carbon Emissions	0.0416	0.0338	0.0376
Resource Consumption	0.1048	0.1143	0.1095
Pollution Emissions	0.1724	0.1456	0.1591
Energy Structure	0.0349	0.038	0.0365
Carbon Productivity	0.0343	0.041	0.0376
Low-Carbon Transportation	0.105	0.1139	0.1094
Low-Carbon Life	0.1094	0.1114	0.1109
Carbon Sink Capacity	0.1088	0.1082	0.1085
Urbanization	0.036	0.0363	0.036
Quality of Life	0.1107	0.1063	0.1084
Economic Development Level	0.0688	0.0798	0.0743

The entropy weights of the indicators were ranked as follows (largest to smallest): pollution emissions > low-carbon life > resource consumption > low-carbon transportation > carbon sink capacity > quality of life > economic development level > amount of carbon emissions > carbon productivity > intensity of carbon emissions > energy structure > urbanization. The importance of each primary indicator in the comprehensive evaluation of low-carbon development was ranked as follows (most to least important): low-carbon production > low-carbon consumption > social and economic development > carbon emissions > low-carbon policy. In addition, several secondary indicators, including pollution emissions, low-carbon life, and resource consumption, were found to be important factors for the evaluation of the level of low-carbon development in cities, as their entropy weights were relatively high. The entropy weights for the indicators of resource consumption, energy structure, carbon productivity, low-carbon transportation, low-carbon life, urbanization, and economic development level increased from 2010 to 2015, while the entropy weights for the indicators of amount of carbon emissions, intensity of carbon emissions, pollution emissions, carbon sink capacity, and quality of life decreased from 2010 to 2015. In the evaluation of the increase in entropy weight, the entropy weight of the indicator for low-carbon transportation increased the most (by approximately 32.38%), while the entropy weight of the indicator for urbanization increased the least (by approximately 0.83%). Among the indicators with entropy weights that decreased from 2010 to 2015, the entropy weight of the indicator for the intensity of carbon emissions decreased the most (by approximately 18.75%), while the entropy weight of the indicator for carbon sink capacity decreased the least (by approximately 0.55%) (Figure 2).

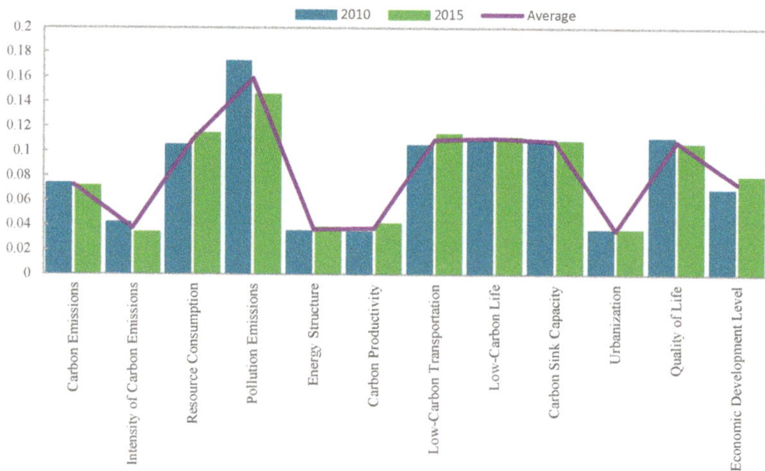

Figure 2. Entropy weights of the evaluation indicators.

4.2. Comprehensive Status Assessment

Overall, after five years of exploration and practice, low-carbon construction achieved remarkable results by the end of 2015. Based on the IPCC Guidelines for Greenhouse Gas Inventories and relevant research results in China and elsewhere, the carbon emission inventory from different routes (energy consumption, agricultural activities, land use change and forestry, and waste disposal) in Xiamen between 2010 and 2015 was roughly calculated (Figure 3).

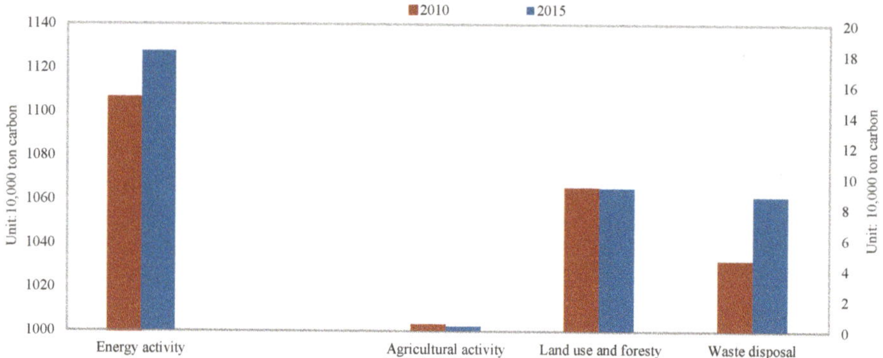

Figure 3. Carbon emissions in Xiamen in 2010 and 2015.

In addition, it should be noted that Xiamen has some particular natural geography and resource conditions. For example, steel, cement, and other industrial materials are not produced locally in Xiamen, but are imported from other places. Therefore, the industrial processes in Xiamen produce very little carbon emission, and hence was excluded from this study. From the perspective of carbon emissions, carbon emissions in Xiamen were mainly from energy consumption, followed by changes in land use and forestry, waste disposal, and agricultural activities. In 2015, after five years of low-carbon development in Xiamen, the carbon emissions generated by agricultural activities, and changes in land use and forestry were significantly reduced to levels lower than those in 2010. In addition, as compared to 2010, although the amount of carbon emission generated by energy consumption and waste disposal in 2015 increased and led to the increase in total carbon emissions, the growth rate of carbon emissions had slowed.

At the same time, compared with 2010, GDP in 2015 increased significantly. According to the decoupling theory [44], the relationship between carbon emissions and economic development showed that GDP increased, and carbon emissions also increased; however, the economic growth rate was higher than the growth rate of carbon emissions. Therefore, while economic growth was achieved, energy consumption was gradually reduced and decoupled from economic growth, to a relative decoupling state (Figure 4). In addition, the intensity of energy consumption for each unit of carbon emissions increased, and the efficiency of resource utilization gradually increased. In addition, pollutant emissions for each unit of carbon emissions decreased, thus improving the quality of the local air and environment to some extent.

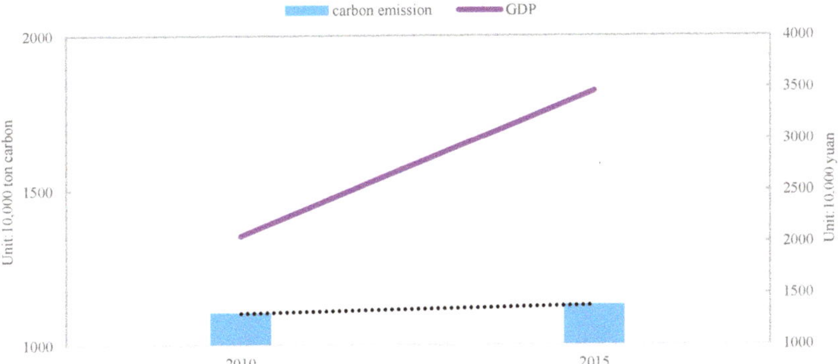

Figure 4. Changes in carbon emissions and GDP.

The values of positive evaluation indicators of low-carbon cities in Xiamen all increased from 2010 to 2015. The indicators of urbanization rate, energy consumption structure, low-carbon transportation, and carbon sink capacity changed the most. The values of the most negative evaluation indicators were reduced, with the exceptions of electricity consumption and waste generation. The increase in electricity consumption and waste generation is closely related to the large influx of residents into the city and the nature of the city itself. In 2015, the indicators of low-carbon development were significantly higher than those in 2010, in terms of low-carbon consumption, low-carbon policy, and social and economic development, while low-carbon development in terms of carbon emissions and low-carbon production require further improvement. However, in general, the comprehensive index of low-carbon development in Xiamen increased from 2010 to 2015, indicating that, after five years of construction and practice, the level of low-carbon development in Xiamen has increased (Figure 5).

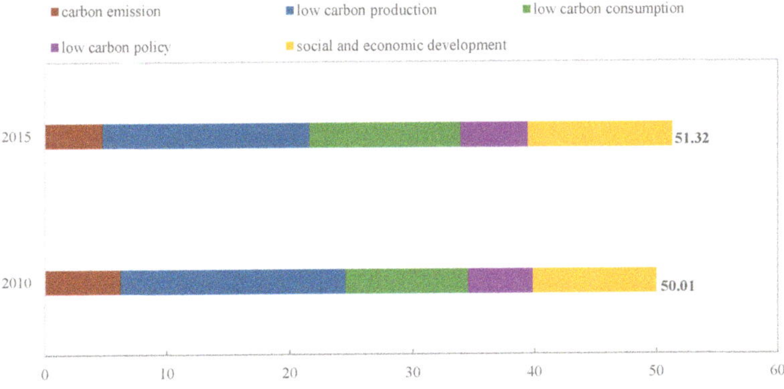

Figure 5. Level of low-carbon development in Xiamen.

5. Discussion

The major innovation of this study was the assessment of representative index systems in China and elsewhere to select a set of indicators for index standardization and construction of an index system for low-carbon cities. By combining the perspectives of three types of index systems, this study avoided the incompleteness of prior index systems constructed from a single perspective and fully integrated the advantages of each perspective, thus allowing us to build a more scientific and reasonable evaluation index system for low-carbon cities. In order to ensure the objectivity of the results, the indicators were all quantitative indicators, which reduced the impact of human subjective factors. At the same time, this approach ensured that the data for the selected indicators were easy to obtain, which made the evaluation protocol realistic and feasible. Another innovation of this research was the selection of an objective weighting method and the entropy weight method, which directly determines the importance of indicators according to changes in indicator data in order to assign weights to the indicators. As a result, fair evaluations can be ensured when different types of cities are compared. In the past, most of the relevant studies adopted qualitative weighting methods, such as the analytic hierarchy process. The shortcoming of this method is that it is too subjective, which makes the results of the evaluation less convincing. Based on the two innovations described above, this study established an evaluation index system for low-carbon cities that consisted of five primary indicators: carbon emissions, low-carbon production, low-carbon consumption, low-carbon policies, and socio-economic development. In addition, evaluations of low-carbon development in cities can be divided into two categories: internal evaluations and external evaluations. Internal evaluations mainly focus on vertical comparison within cities and evaluate the low-carbon development of cities by selecting indicators relevant to five aspects: carbon emissions, low-carbon production, low-carbon consumption, low-carbon policies, and social and economic development. In this way, low-carbon development, related to all the different aspects of cities, can be evaluated. External evaluations mainly involve horizontal comparisons of low-carbon development among all provincial capital cities and municipalities directly under the central government of China. By comparing the overall low-carbon development among these cities in terms of indicators related to carbon emissions, a comprehensive index ranking can be determined to reflect the real level of low-carbon development in these cities. At the same time, the relevant government administrations can also scientifically and reasonably evaluate the level of low-carbon development among similar cities using external evaluation, so that the cities can develop various improvement measures that are conducive to low-carbon development (Figure 6).

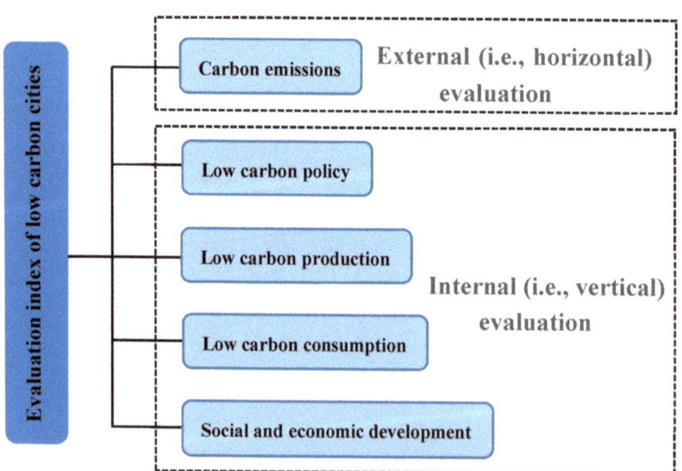

Figure 6. Method for selecting low-carbon evaluation index.

There are some shortcomings of this study. Due to limited data availability, some indicators were discarded despite their usefulness for evaluating low-carbon cities. For example, indicators for low-carbon buildings (proportion of energy-saving buildings, and energy consumption per unit building area), low-carbon technologies (capture and storage ratio of carbon dioxide), and low-carbon policies (completeness of low-carbon policies and regulations, public satisfaction with low-carbon cities) were excluded from this study. At the same time, when the carbon emissions of Xiamen were calculated according to the IPCC Guidelines for Greenhouse Gas Inventories, carbon emissions generated by industrial production were ignored because Xiamen has a particular natural geography and resource conditions. For example, steel, cement, glass, and other industrial products with high carbon emissions are not produced locally in Xiamen, but are instead imported from other places.

In future studies, we will explore a weighting method that utilizes qualitative and quantitative indicators, try to find an alternative for necessary indicators with limited data availability, and study the level of low-carbon development in cities with a longer time scale, while simultaneously considering the spatial scale of the analysis. At the same time, as this research is aimed at the standardization of indicators, we need to refer to the relevant standards of domestic and foreign cities to set the benchmark indicators. It is necessary to expand to a wider range of cities for comparison, such as at the global, regional, and national level.

6. Conclusions

Based on the results of previous studies, this study integrated three perspectives used to construct evaluation index systems for low-carbon cities. By extracting common indicators and adjusting individual indicators, a standardized evaluation index system for low-carbon cities was established. The established evaluation index system can be used to horizontally compare the levels of low-carbon development among different cities, as well as to evaluate the levels of low-carbon development in specific sectors or industries within a city. The content of these indicators reflects current problems and routes to improve low-carbon development. The newly developed index system uses a balanced configuration of positive and negative indicators, including five primary indicators for carbon emissions, low-carbon production, low-carbon consumption, low-carbon policies, and social and economic development. In addition, all indicators in the index system are quantitative indicators, and the relevant data can be obtained directly or calculated from statistical yearbooks. Therefore, in the future, quantitative comparisons of the levels of low-carbon development among different cities can avoid the uncertainty caused by human subjective factors, and the results of such evaluations will be more objective and comparable than the results of evaluations conducted using current index systems. Moreover, in comparison with current index systems, the newly developed index system can be used in a broader range of applications.

Using the entropy weight method, low-carbon development in Xiamen from 2010 to 2015 was evaluated quantitatively and comprehensively. Pollution emissions had the greatest impact on low-carbon development in Xiamen, while urbanization had the least impact. In terms of the comprehensive index of low-carbon development in Xiamen, the level of low-carbon development in 2015 was better than that in 2010 due to efforts at promoting energy conservation, emission reduction, and green development. In addition, in terms of carbon emissions, the growth rate of carbon emissions from energy activities, agricultural activities, changes in land use and forestry, and waste treatment slowed significantly in 2015 compared with that in 2010. In addition, the amount of carbon emission generated from agricultural activities, and changes in land use and forestry in 2015 was less than that in 2010. These changes were closely related to efforts at increasing afforestation and improving forest resource management. At the same time, the growth rate of carbon emissions was slower than the growth rate of economic development, and these rates were decoupled. Although the growth rate of carbon emissions slowed from 2010 to 2015, the total amount of carbon emission in 2015 was higher than that in 2010, mainly because energy activities were still the main source of urban carbon emissions. Furthermore, the amount of carbon emission generated by waste disposal was still high, mainly because of Xiamen's large population, which generated a large amount of household waste. Moreover, most of the waste produced in Xiamen was destroyed by incineration, which also acted as an important contributor to carbon emissions.

Author Contributions: L.S. and X.X. conceived and designed the structure and case study of paper, and they processed data and wrote original draft; W.Z. collected data and information; L.G. did the statistics work and supervised research.

Acknowledgments: This research was supported by the National Key Research & Development Program of China (2018YFC0704703, 2017YFF0207302). We would like to thank Jianyi Lin at Institute of Urban Environment, Chinese Academy of Sciences for his advice on the study.

Conflicts of Interest: The authors declare no conflict of interest. The founding sponsors had no role in the design of the study; in the collection, analyses, or interpretation of data; in the writing of the manuscript, and in the decision to publish the results.

References

1. Tu, C.; Li, F.D. Responses of greenhouse gas fluxes to experimental warming in wheat season under conventional tillage and no-tillage fields. *J. Environ. Sci.* **2017**, *54*, 314–327. [CrossRef] [PubMed]

2. Zhou, Y.; Hao, F.H.; Meng, W.; Fu, J.F. Scenario analysis of energy-based low-carbon development in China. *J. Environ. Sci.* **2014**, *26*, 1631–1640. [CrossRef] [PubMed]

3. BP (British Petroleum). BP World Energy Statistics. Available online: https://www.bp.com/en/global/corporate/media/speeches/bp-statistical-review-of-world-energy-2017.html,2017-07-13/2018-08-15 (accessed on 15 August 2018).

4. Jiang, J.J.; Ye, B.; Ma, X.M. The construction of Shenzhen's carbon emission trading scheme. *Energy Policy* **2014**, *75*, 17–21. [CrossRef]

5. UN (United Nations). World Urbanization Prospects the United Nations, New York. Available online: https://esa.un.org/unpd/wup/publications/files/wup2014-highlights,2014-07-21/2018-09-20 (accessed on 15 October 2018).

6. CG (Chinese Government). Enhanced Actions on Climate Change: China's Intended Nationally Determined Contributions. Available online: http://www.china.org.cn/chinese/2015-07/01/content_35953590.htm, 2015-07-01/2018-09-20 (accessed on 20 September 2018).

7. Du, H.B.; Chen, Z.N.; Mao, G.Z.; Li, R.Y.M.; Cai, L.H. A spatio-temporal analysis of low-carbon development in China's 30provinces: A perspective on the maximum flux principle. *Ecol. Indic.* **2018**, *90*, 54–64. (In Chinese) [CrossRef]

8. Azizalrahman, H.; Hasyimi, V. Towards a generic multi-criteria evaluation model for low carbon cities. *Sustain. Cities Soc.* **2018**, *39*, 275–282. [CrossRef]

9. Hak, M.; Matsuoka, Y.; Gomi, K. A qualitative and quantitative design of low-carbon development in Cambodia: Energy policy. *Energy Policy* **2017**, *100*, 237–251. [CrossRef]

10. Tan, S.T.; Yang, J.; Yan, J.Y. Development of the Low-carbon City Indicator (LCCI)Framework. *Energy Procedia* **2015**, *75*, 2516–2522. [CrossRef]

11. Kalnins, S.N.; Blumberga, D.; Gusca, J. Combined methodology to evaluate transition to low-carbon society. *Energy Procedia* **2015**, *72*, 11–18. [CrossRef]

12. Zhou, N.; He, G.; Williams, C.; Fridley, D. ELITE cities: A low-carbon eco-city evaluation tool for China. *Ecol. Indic.* **2015**, *48*, 448–456. [CrossRef]

13. Lu, Q. DGNB of Germany-the World's Second Generation Green Building Assessment System. Available online: https://wenku.baidu.com/view/d31e3358312b3169a451a4ca.html,2012-08-22/2018-09-20 (accessed on 20 September 2018).

14. Pamlin, D. Low Carbon City Index. Available online: http://www.pamlin.net.2009 (accessed on 15 January 2015).

15. Zhang, L.P.; Zhou, P. A non-compensatory composite indicator approach to assessing low-carbon performance. *Eur. J. Oper. Res.* **2018**, *270*, 352–361. [CrossRef]

16. Yang, X.; Wang, X.C.; Zhou, Z.Y. Development path of Chinese low-carbon cities based on index evaluation. *Adv. Clim. Chang. Res.* **2018**, *9*, 144–153. [CrossRef]

17. Ohshita, S.; Zhang, J.; Yang, L.; Hu, M.; Khanna, N.; Fridley, D.; Liu, S.; Chen, L.Y.; Li, A.; Sun, M.; et al. China Green Low-Carbon City Index. Available online: https://escholarship.org/uc/item/6m30b8$\times$7,2017-08-15/2018-09-20 (accessed on 20 September 2018).
18. Price, L.; Zhou, N.; Fridley, D. Development of a low-carbon indicator system for China. *Habitat Int.* **2013**, *37*, 4–21. [CrossRef]
19. Fu, Y.; Liu, Y.J.; Wang, Y.L. Research on the evaluation method and support system of low carbon cities. China Population. *Resour. Environ.* **2010**, *20*, 44–47. (In Chinese)
20. Hua, J.; Ren, J. Research on the evaluation of low carbon cities based on ANP. *Sci. Technol. Econ.* **2011**, *24*, 101–105. (In Chinese)
21. Xin, L. Construction of evaluation index system for low carbon cities. *Stat. Decis.* **2011**, *7*, 78–80. (In Chinese)
22. Zhu, J.; Liu, X.M.; Zhang, Y. Establishment of an index system for the evaluation of low carbon city construction in China. *Ecol. Econ.* **2017**, *33*, 52–56. (In Chinese)
23. Shao, C.F.; Ju, M.T. Research on the index systems of low carbon cities based on the DPSIR model. *Ecol. Econ.* **2010**, *10*, 95–99. (In Chinese)
24. Zhang, X.P.; Liu, J.; Fang, T. Evaluation of low carbon city development in Lanzhou based on the DPSIR model. *J. Northwest Normal Univ. (Nat. Sci. Ed.)* **2012**, *48*, 112–115. (In Chinese)
25. Zhang, L.; Chen, K.L.; Cao, S.K. Construction of an evaluation index system of low carbon cities from the perspective of carbon source/sink. *Energy Environ. Prot.* **2011**, *25*, 8–11. (In Chinese)
26. Chu, C.L.; Ju, M.T.; Wang, Y.N.; Wang, Y.S. Discussion on the ideas and technical framework for the planning of low carbon city development in China. *Ecol. Econ.* **2011**, 45–48. (In Chinese)
27. Media Alliance of China Low-Carbon Economy. China's Low Carbon City Evaluation System. Available online: http://www.chinanews.com/ny/2011/01-20/2799909.html,2011-01-20/2018-09-20 (accessed on 20 September 2018). (In Chinese)
28. Chatham House; Chinese Academy of Social Sciences; Energy Research Institute of National Development and Reform Commission. Plan for Low Carbon Development in Jilin City. Available online: http://www.doc88.com/p-382770423918.html.2010-03-31/2018-09-20 (accessed on 20 September 2018). (In Chinese)
29. Zhou, G.H.; Singh, J.; Wu, J.C.; Sinha, R.; Laurenti, R.; Frostell, B. Evaluating low-carbon city initiatives from the DPSIR framework perspective. *Habitat Int.* **2015**, *50*, 289–299. [CrossRef]
30. Du, D.; Wang, T. Comprehensive evaluation of the improvement and development of the evaluation index system for low-carbon cities. *Chin. J. Environ. Manag.* **2011**, *3*, 8–11, 14. (In Chinese)
31. Tan, S.T.; Yang, J.; Yan, J.Y.; Lee, C.; Hashim, H.; Chen, B. A holistic low carbon city indicator framework for sustainable development. *Appl. Energy* **2017**, *185*, 1919–1930. [CrossRef]
32. Tan, Q. Construction of a low carbon city evaluation index system: A case study of dynamic comparison between Nanjing and Shanghai. *Ecol. Econ.* **2011**, *12*, 81–84, 96.
33. Li, B.H.; Xu, L. Measurement of the development level of low carbon cities and its countermeasures: A case study of Changsha, Zhuzhou and Xiangtan. *J. Anhui Agric. Sci.* **2011**, *39*, 1180–1183. (In Chinese)
34. Lian, Y.M. Urban value and the evaluation index system of low carbon cities. *Urban Probl.* **2012**, *1*, 15–21. (In Chinese)
35. International Organization for Standardization. Sustainable Development of Communities—Indicators for City Services and Quality of Life. Available online: https://www.iso.org/standard/62436.html.2014-05-15/2018-09-20 (accessed on 20 September 2018).
36. BNU (Beijing Normal University). *China Green Development Index*; Beijing Normal University Press: Beijing, China, 2016. (In Chinese)
37. Yi, D.J. Evaluation of Low Carbon Cities in Central Provincial Capital. Master's Thesis, Central South University, Changsha, China, 2010. (In Chinese)
38. Zhang, X.L. Research on TOPSIS-Based Evaluation of Low Carbon Cities in China. Master's Thesis, Jilin University, Changchun, China, 2017. (In Chinese)
39. Yang, Y.F. Study on the Evaluation System of Low Carbon City Development-Taking Beijing as an Example. *Anhui Agric. Sci.* **2012**, *40*, 344–351.
40. Ma, L.; Liu, X.G.; Liu, Z.W. Study on the Evaluation Index System and Model Construction of Low-carbon Cities. *J. Jinan Univ. (Soc. Sci. Ed.)* **2014**, *24*, 55–59.
41. Wang, Y.Z.; Zhou, Y.Y.; Deng, X.Y. Construction and Empirical Analysis of Evaluation Index System for Low-carbon Cities. *Stat. Sci. Pract.* **2011**, *1*, 48–50. (In Chinese)

42. IPCC. IPCC Guidelines for National Greenhouse Gas Inventories. Available online: https://wenku.baidu. com/view/da92d91dc5da50e2524d7fcd.html,2011-11-21/2018-09-20 (accessed on 20 September 2018).
43. Cai, B.F. Research on the list of urban greenhouse gas in China. China Population. *Resour. Environ.* **2012**, *22*, 21–27. (In Chinese)
44. Zhao, R.Q. Study on the Carbon Cycle of Urban ECO-economic System and ITS Land Regulation Mechanism. Ph.D. Thesis, Nanjing University, Nanjing, China, 2011. (In Chinese)
45. Shannon, C.E. A mathematical theory of communications. *Bell Syst. Tech. J.* **1948**, *27*, 623–666. [CrossRef]
46. Yang, Z.H.; Zhang, C.L.; Ge, L.; Gong, D.C.; Gu, Y.F.; Huang, T.h. Comprehensive fuzzy evaluation of insulator flashover state based on an entropy weight method. *Electr. Power Autom. Equip.* **2014**, *34*, 90–94.

Article

Qualifying the Sustainability of Novel Designs and Existing Solutions for Post-Disaster and Post-Conflict Sheltering

Lara Alshawawreh [1], Francesco Pomponi [2,*], Bernardino D'Amico [2], Susan Snaddon [3] and Peter Guthrie [4]

[1]   Faculty of Engineering, Mutah University, Karak 61710, Jordan; laraalshawawreh@gmail.com
[2]   REBEL (Resource Efficient Built Environment Lab), School of Engineering and the Built Environment, Edinburgh Napier University, 10 Colinton Road, Edinburgh EH10 5DT, UK; b.damico@napier.ac.uk
[3]   Independent Built Environment Consultant, Nairobi 00200, Kenya; susansnaddon@gmail.com
[4]   Department of Engineering, University of Cambridge, Trumpington Street, Cambridge CB2 1PZ, UK; pmg31@cam.ac.uk
*   Correspondence: f.pomponi@napier.ac.uk

Received: 27 November 2019; Accepted: 19 January 2020; Published: 24 January 2020

Abstract: During the course of 2018, 70.8 million people globally were forcibly displaced due to natural disasters and conflicts—a staggering increase of 2.9 million people compared to the previous year's figure. Displaced people cluster in refugee camps which have very often the scale of a medium-sized city. Post-disaster and post-conflict (PDPC) sheltering therefore represents a vitally important element for both the short- and long-term wellbeing of the displaced. However, the constrained environment which dominates PDPC sheltering often results in a lack of consideration of sustainability dimensions. Neglecting sustainability has severe practical consequences on both people and the environment, and in the long run it also incurs higher costs. It is therefore imperative to quickly transfer to PDPC sheltering where sustainability considerations are a key element of the design and decision-making processes. To facilitate such transition, this article reviews both 'existing solutions' and 'novel designs' for PDPC sheltering against the three pillars of sustainability. Both clusters are systematically categorized, and pros and cons of solutions and designs are identified. This provides an overview of the attempts made so far in different contexts, and it highlights what worked and what did not. This article represents a stepping-stone for future work in this area, to both facilitate and accelerate the transition to sustainable sheltering.

Keywords: city; post-disaster shelter; post-conflict shelter; transitional shelter; sustainable sheltering; emergency sheltering; refugees

---

1. Introduction

According to the United Nations High Commissioner for Refugees [1], 70.8 million people around the world were forcibly displaced during 2018 due to natural disasters and conflicts, exceeding the previous year's numbers by 2.3 million people. Displacement is a complex challenge, which if not properly addressed could fuel existing tensions and create new conflicts [2]. This is something to consider when planning for sheltering, as is accounting for the local culture and the context of each individual situation. In post-disaster and post-conflict (PDPC) situations, it usually takes two to fifteen years to resolve land rights, which affects the reconstruction of damaged homes [3]. Therefore, providing shelters in the initial stages after disasters and conflicts is critical to ensure adequate levels of safety, security, protection and community health [4]. The UNHCR, IFRC and their operational partners, who are responsible for providing shelters to the affected population, usually have limited time and funding to propose shelter solutions when a disaster occurs, which affects the quality of the shelters. There have been attempts by companies and researchers to design shelter solutions, but usually they prioritize the transportability and rapid deployment of the shelters. They rarely consider the social and cultural factors or the visual, acoustic and thermal performance [5].

Johnson [6] identified some of the main challenges facing PDPC shelters as high costs, delivery delays, remote and adverse locations, and poor design—generally unsuitable for both the users' culture and the local climate conditions. It also seems that the existing literature focuses more on solutions from outside the area involved, with little or no acknowledgement of the solutions found informally by the affected people and local communities [7].

In the global movement to urgently transfer into a more sustainable way of living, the humanitarian sector has been given insufficient attention. In the past, it was seen as an indulgence to consider the environment in PDPC responses, due to the significant size of the affected population and the crisis intensity. The fact that environmental damage contributes to natural disasters was surprisingly neglected [8]. The environmental impact of the aid shelters has been highlighted as a clear knowledge gap by Ramboll and Save the Children [8], and the need for further research is amongst their recommendations. The same gap was highlighted by Albadra, Coley and Hart [9] as their literature survey showed that in the past 38 years, only 60 academic papers have been published on the subject of 'emergency or temporary shelters', and only nine of them addressed the shelters' life cycle sustainability or environmental impacts.

The importance of sustainable development is explained in the definition of the Brundtland Report as "meeting the needs of the present without compromising the ability of future generations to meet their own needs" [10]. In the PDPC scenarios, Potangaroa [11] argues that building informal shelters could have adverse effects on the environment and deplete resources, which might also raise the cost of materials and labor, and potentially lower the quality of life by ignoring the previous social structures and cultural spaces. However, Potangaroa [11] adds that designing sustainable shelters would avoid these issues and have long-lasting positive impacts. This article aims therefore to contribute to ongoing global efforts to design more sustainable shelters. To do so, an important and useful first step is mapping the range of 'existing solutions' and 'novel designs' to identify their strengths and weaknesses. Since sustainability is a holistic concept that involves not only the environmental dimension, but also economic and social aspects [12], these three elements constitute the key parameters against which the review presented here is carried out. In the next section, the existing literature is reviewed and the adopted shelter terminologies and typologies in this paper are clarified. The section also discusses the main arguments regarding the three pillars of sustainability. In Section 3, the methodology that was adopted in this paper is clarified and the chosen case studies are analyzed in comparative tables that show the pros and cons of each case. The results of the comparisons are presented in Section 4, and discussed in Section 5. Lastly, Section 6 concludes the article and sets the challenges and recommendations for future research.

2. Literature Review

In PDPC situations, decision makers generally prefer to direct most of the money and effort into the reconstruction phase rather than the initial relief sheltering. However, previous cases teach us that such an approach may have major issues. The shelter response after the earthquakes in Ardabil and Lorestan Provinces in Iran, for instance, is an example where unexpected events delayed the reconstruction process, resulting in thousands of people living in emergency tents for up to two years, remaining unprotected from the harsh weather [13]. Hadafi and Fallahi [13] also argued that if people were consulted in the first place about how to deal with the post-emergency, they might have chosen a different approach, and therefore the adverse effects would have been lessened. In order to address the literature on PDPC shelters, an agreement on terminology is necessary. Therefore, a reflection on the history of the terms and the confusion in their usage is presented, followed by the common shelter categorization, prior to reviewing key social, environmental and economic aspects.

According to Chang et al. [14], the availability of resources in PDPC situations is very challenging and could cause environmental and economic obstacles. Moreover, the same study concludes that stakeholders' collaboration and advanced planning would allow the market, donors and governments to successfully manage the available resources. This is explained in more detail in the Humanitarian Emergency Response Review [15] as it describes the gap between the humanitarian challenge and the world's ability to cope as a lost race. The report referred to two major challenges in dealing with humanitarian crisis; first, the global economic crisis and second, the rising security threat that affects the performance of the humanitarian workers in the field and makes the assessment of affected people in PDPC situations a harder mission. However, there is an argument that the real reason behind the failure of shelters is the undeveloped understanding of the topic by the working institutions despite their progress in the field [16]. Moreover, few external agencies enter the field of shelters, due to its complexity and high consumption of time, cash and energy [16].

2.1. Shelter Terminologies

There are no agreed terminologies regarding sheltering, and the existing terms are usually confusing. United Nations Disaster Relief Organization [17] suggested eight phases of shelter provision: tents, imported designs and units, standard designs incorporating indigenous materials, temporary housing, the distribution of materials, core housing, hazard-resistant housing, and accelerating reconstruction of permanent housing. Thirteen years later, Quarantelli [18] proposed a different shelter categorization based on the life span of the shelters and people's behavior, which included four stages: emergency sheltering, temporary sheltering, temporary housing and permanent housing. Quarantelli [18] explained that these terms are the ideal categories and do not necessary reflect the actual reality. The unique addition of this study was the terms' differentiations he proposed. Specifically, he distinguishes between *emergency* and *temporary* shelters—mainly in the behavioral aspects, where the temporary responds to where and how the daily activities are held. He also distinguishes between *sheltering* and *housing*, where in housing the users resume their routine household responsibilities and activities, which does not happen in sheltering. Finally, he distinguishes between *temporary* and *permanent* housing, where the users in the latter live in a repaired, rebuilt or new permanent physical structure. However in reality, some temporary housing was never vacated and was transformed into permanent homes [18]. Barakat [19] proposed different definitions for *shelter* and *housing*. He defines the shelter as a structure intended for temporary use in spite of the real duration of its usage. Housing instead provides either a permanent solution or a solution that supports the affected communities until they can rebuild their own homes. Hadafi & Fallahi [13] clarify that housing is not only about the physical shelter structure but instead a system that is concerned with people's social, psychological and spiritual needs. More recently, the International Federation of Red Cross and Red Crescent Societies (IFRC) suggested six shelter duration levels of what they call 'approaches' instead of the typical 'response phases'. The approaches, which mainly depend on the context of each case, are: emergency shelter, temporary shelter, transitional shelter, progressive shelters, core shelters, and permanent housing [20].

Most 'novel designs' and 'existing solutions' claim to fall under the transitional shelter category, which is inaccurate in most cases. Such a misattribution is due to two major misconceptions: (1) the view that transitional shelter is a product, and (2) the use to describe approaches to permanent construction [3]. The transitional shelter is the incremental process that provides sheltering to the affected families while they are seeking to maintain other recovery options. This happens through its five characteristics; upgradability, reusability, ability to be relocated, ability to be resold and recyclability [3]. Transitional shelters usually represent a first step prior to relocation to more durable sheltering solutions and are usually built on a temporary site [20]. Since the studied shelters in this paper are classified into 'novel designs' and 'existing solutions' despite their phase or approach, this paper does not adopt any terminologies for the phases/approaches. In addition to the confusion in the phases' terminologies, the general term used to refer to the sheltering response is not agreed on, but the terms 'emergency shelters' and 'temporary shelters' are commonly used amongst scholars. However, since the previous two terms are used to describe specific approaches/phases in some categorizations including the IFRC [20], 'PDPC shelters' is proposed as an umbrella term in this article.

2.2. Shelter Typologies

Scholarly classification of shelters is diverse. Albadra, Coley and Hart [9] categorized the shelters in terms of their manufacturing approach or location into 'transportable shelters' and 'built on-site shelters'. They clarify that transportable shelters include any shelter that is manufactured off-site and then shipped to the intended location. This category covers both basic shelters such as tents, and more developed flat-packed solutions. Conversely, the built on-site shelters are usually constructed using locally available materials. In most cases, the beneficiaries are provided with tool kits and training sessions to allow them build their own shelters. A similar categorization was created by Felix, Branco and Feio [21]. They grouped the shelters based on their readiness level into 'ready-made units' and 'kit supplies'. The ready-made units are fully constructed in a factory environment and transported to the location as one item. They could be divided into separate but somewhat large parts to be assembled on site. Kit supplies instead solve the problem of heavy transport systems by producing smaller elements that can be erected by local people on-site.

The challenge with the previous two categorizations is twofold. Firstly, there will be a confusion in when to consider the parts as a ready-shelter that is divided into pieces (transportable) or parts of a kit (built on-site). Secondly, the applicability of the shelters is not considered, as many ideas are logical in theory but have never been tested in real PDPC situations.

Quaglia, Dascanio and Thrall [22] analyzed the existing US military solutions in order to present their origami-inspired proposals for what they called 'rapidly deployable shelters'. They categorized the military shelters depending on the wall attributes into 'non-expandable rigid wall shelters', 'expandable rigid wall shelters', and 'soft wall shelters'. Analyzing the military shelter solutions alongside the PDPC shelters is problematic, as the context and needs of the military and shelters for PDPC persons differ significantly. This research instead adopts a categorization method based on the historical application of the shelters, which will be discussed in the methodology section.

2.3. Social, Environmental and Economic Aspects

The main sustainability challenges facing current PDPC shelters are the lack of cultural adequacy, economic viability and negative environmental impacts [21]. The recognition of the importance of users' participation in order to have culturally sensitive designs has been acknowledged by NGOs, policy makers and scholars, specifically in reconstruction. Thirty-six years ago, the UNDRO [17] concluded that the key to success in reconstruction is the local community's participation. Opdyke, Javernick-Will and Koschmann [23] found that early user involvement in PDPC shelter projects, support the resilience and the sustainability of the project outcome. The local input could empower the affected people, encourage the social connectivity and promote solidarity between the beneficiaries themselves [24]. Although it is not always easy to measure the success of a shelter project, but user satisfaction and shelter safety are major indications of that success [24], which could be evaluated through surveys and observations such as the shelter assessment that was done in Zaatari camp [25] and Azraq camp [26]. However, sometimes the lack of users involvement leads to an obvious project failure when the assistance is not accepted by the people, such as the refusal of using the steel caravans in Gaza [27]. It should be noted, however, that engagement the affected people might be in practice challenging in several cases where there is a great urgency to act, the people are in trauma, and potentially not well placed to be consulted, which would also lead to further delay in the aftermath of an emergency. The case study of Al-burjan village in Lebanon is an example, where the main lesson learnt was that reconstruction must be culturally rooted, i.e., responds to the cultural needs and the perceptions of the local people, which could only be met by involving the affected people in the rebuilding process from its early stages [28]. Barakat and Zyck [29] proposed a 'hybrid approach' in reconstruction that combines the 'owner driven' and 'contractor driven' existing approaches that are used in Southern Lebanon. The purpose was to ensure the structural integrity of the house through a contractor constructing the foundation and the frame, while at the same time allowing the owners to design the layout. Not only does the cultural inadequacy in designing shelters result in uncomfortable living conditions, but it also causes serious social problems within the communities. However, little is known about how and when to apply the principles of participation in the PDPC sheltering design. Early participation provides better and more satisfying design results. It also empowers the affected population and allows them to be active again in the society, instead of being perceived as passive help-recipients. Unfortunately, this dimension has been neglected in existing literature [30]. The shelters in the Jordanian Syrian-camps are examples of cultural inadequacy which was obvious through the adjustments made by the residents to their own shelters, such as adding dividers to the one-room design in order to separate the sleeping areas for family members of different age and gender [31]. The Sphere Association [32] emphasizes the importance of considering the needs, preferences and habits of various age, gender and disability groups.

In regard to the environmental perspective, Felix, Branco and Feio [21] recommend planning flexible shelters that could be reused after the initial purpose or period ends. A study by Escamilla and Habert [33] argues that while global materials could provide structures with high embodied energy and high resistance levels against natural hazards, local materials need extra attention on the structural details to withstand the possible hazards and therefore could increase the economic and environmental costs. Therefore, the study concluded that sustainable shelter solutions can be produced using either global or local construction materials, as global materials will most likely provide better technical performance while the local materials will likely lower both costs and environmental impact. They also found that cost and environmental impact do not necessarily affect the technical performance of shelters. Escamilla and Habert [33] consider the global materials as "industrialized and engineered construction materials like concrete and steel", while consider the local materials as those used in "traditional and vernacular architecture, like bamboo, earth/soil and wood". In a follow-on study, Celentano et al. [34] found that the material supply (local or global) is the main factor affecting the speed of the construction in the construction technology scale, and noted that using local materials decreases the cost but increases the construction time, while the use of industrialized materials does the opposite. However, when focusing on the scale of the shelter unit, they found that the roof's complexity is the main factor affecting the speed and not the source of the materials. Therefore, they suggest using local materials with a small input of industrialized materials to increase the speed with no noticeable impact on costs. The International Organization for Migration (IOM) recommends the selection of culturally appropriate materials (i.e., local materials that are already used for traditional and vernacular architecture within the existing culture), as it helps protecting the natural resources, and reflects the local expertise in resource management—which consequently, will reduce the shelters carbon footprint through minimizing the energy consumption and pollution [3].

The shelter's total cost usually includes the cost of materials, transportation, construction work and the workforce, but excludes the cost of the camps' infrastructure, which results in having inaccurate comparisons. Additionally, the cost of waste management that results from materials' manufacturing should be considered [35]. The intended short lifespan of PDPC shelters makes the investment in their quality appear inefficient as it could result in them costing more than permanent housings [21]. This however generally proves false for two reasons: shelters usually stay in their place, and are occupied, for much longer than initially anticipated (1) and considering only the initial costs when comparing solutions is short-sighted, as the operational costs differ widely when a well-designed shelter is used for a long time (2). This calls for a greater adoption of life cycle thinking in future PDPC sheltering solutions as the missing link between design and sustainability.

3. Methods

Building on the literature review, desk-based research was conducted in order to identify the global and regional solutions for PDPC sheltering as this information is not available within existing academic literature. These designs have been categorized according to their historical application into 'novel designs' and 'existing solutions'. Novel designs are defined in this research as shelter designs developed by researchers or companies but not necessarily ever used. Existing solutions are instead those applied in the field as a response to PDPC shelter needs, most commonly by UNHCR, IFRC or their partners. In order to clarify the difference between the two categories, any shelter that was prototyped and used by more than 100 refugee families is categorized as 'existing', while other shelter designs that were never prototyped or were prototyped and used by less than 100 refugee families, are categorized as 'novel designs'. While the information about 'novel designs' were collected from their official web pages and magazine articles, the data for the 'existing shelters' were sourced from the organization documents, mainly Shelter Projects [36]. Two considerations influenced the choice of the cases; the quantity of available information and the variety of shelters—by material and geographic spread.

The shelters in both clusters have been investigated against three dimensions (social, environmental and economic) from a qualitative perspective. This approach is well accepted in the sustainability field [37,38], adopted in UN-endorsed research [39] and across sectors, spanning from the circular economy [40] to oil and gas operations [41]. Arguably, a quantitative approach could have provided richer information but we fully agree with Janoušková and colleagues [42]: sustainability indicators are useful but they must also be relevant. For instance, Matard et al. [43], recently analysed embodied energy and embodied greenhouse gas (GHG) emissions of a 81 refugee shelters worldwide. This is a significant research endeavor but at a closer look the relevance of their findings comes into question since "lists of materials were inferred from pictures and details mentioned in the narratives" [43] (p. 33) and their values are fully based on an old version of the ICE database [44], which is specific to the UK construction industry and not reviewed. Why should it therefore produce results that are appropriate for and relevant to shelters in developing countries? Using numbers to produce other numbers is relatively easy, ensuring relevance in the process very less so. For this reason, and due to the lack of data about the shelter designs that would allow to quantify the sustainability dimensions meaningfully, we resisted the temptation of a quantitative approach in this article.

Following the qualitative approach just explained, pros and cons (as emerged from the literature review and outlines in Section 2.3) have been noted for each dimension for each shelter solution. The following two tables illustrate the examined 24 cases with the comparisons; 12 cases were categorized as 'novel designs' within Table 1, while the other 12 cases were categorized as 'existing solutions' and are shown in Table 2. Tables with full details are available in the Supplementary Material. Specifically, Table S1 for the 'novel designs', and Table S2 for the examined 'existing solutions'. However, it is important to note that the information in the tables is based on how the projects are reported in the references used and the availability of information and therefore could be limited or not verified. Again, a degree of qualitative interpretation was needed to classify the indicators into pros and cons. This is mainly obvious in the environmental sustainability, which is a concept that cannot be classified in terms of absolute pros and cons, as the paths leading to it will differ in each country or sector [45]. Additionally, if environmental sustainability is to be measured it must first be quantified. We therefore suggest the use of environmental impact assessment tools, such as Life Cycle Assessment (LCA), in order to inform environmentally-sustainable decisions. The work of Matard and colleagues [43] is an excellent first step in this direction, and it is hoped that future works will use appropriate datasets and reliable quantities, and be expanded beyond the cradle-to-gate boundary.

For the scope of this article, based on the available information, and for the sake of our categorization, we have clustered, for instance, on the pros side of environmental sustainability examples where local materials were reported to be used (materials used in the traditional and vernacular architecture) and on the cons side cases where global materials (industrialized) were used. This choice reflects their lower reliance on fossil fuels, lower energy- and carbon-intensive supply chains, and the likely availability within shorter distances from the site where they are employed, thus reducing global transportation impacts. There remain however instances where a categorization would be misleading. This is the case of using perlite in the Tentative Concept design, for example, which is on the one hand a natural material and, on the other, a possible cause of rhinitis and pneumonia [46]. A similar consideration could also apply to stone wool.

Table 1. Novel designs sustainability comparison.

	Shelter Solution	Social Sustainability		Environmental Sustainability		Economic Sustainability		Ref
		Pros (+)	Cons (−)	Pros (+)	Cons (−)	Pros (+)	Cons (−)	
1	Conrad Gargett's	-Flexible -No mechanical fixings	-Does not consider SN -One room -No private T&K	-Use of wood	-Use of plastic		'Unknown cost'	[47,48]
2	Exo stackable shelter	-Easily deployed -No tools needed -Can attach multi units	-Does not consider SN -One room -No private T&K	-Use of wood -LED light display -Recyclable	-Use of Aluminum -Steel in floor		Unaffordable	[49–51]
3	U-dome	-Easily deployed -Can incorporate LM	-Does not consider SN -One room -Small size -No private T&K	-Compatible to RES	-Use of plastic -Use of Nylon		Above average	[52–54]
4	TranShel	-Easily deployed -Expandable -Possibility of LM	-Does not consider SN -One room -Small size -Low roof height -No private T&K	-Reusable -No off gassing -Recyclable -Possibility of LM	-Use of plastic		Above average	[55,56]
5	Concrete Canvas shelter	-Various sizes -Easily deployed	-Does not consider SN -One room -No private T&K	-Durable -Covered by earth	-Use of concrete -Plastic inner -Vehicle needed		Unaffordable	[57,58]
6	The Liina Transitional Modular Shelter	-Easily deployed -Various rooms -Private K	-Does not consider SN -Small size -No private T	-Use of wood -Insulated panels -Durable	-Use of Nylon		'Unknown cost'	[59,60]
7	The Pallet House	-Easily deployed -Adaptable -LM (P)	-Depends on the availability of materials -No private T&K	-Use of wood -Wood/straw roof (P) -Possibility of LM	-CS roof (P)	Below average (Basic material)		[61]
8	Life shelter	-Easily deployed -Adaptable -LM (P) -Durable	-Does not consider SN -One room -Small size -No private T&K	-Stone wool insulation -Durable -Reusable	-Stone wool insulation -Use of steel -Cement cladding roof	Below average (For large quantities)		[62,63]
9	Rapid Deployment Module (RDM)	-Easily deployed -Integrated floor	-Does not consider SN -One room -Small size -No private T&K	-Passive cooling and heating -Reuse shipping box -Durable	-Unknown walls materials -Questionable TC		Unaffordable	[64–66]
10	Tentative Concept	-Raised floor	-Small size -No private T&K	-Use of fiberglass -Use of textile with Pe -Collects water on roof	-No TC -Use of Pe		'Unknown cost'	[67,68]
11	Hex house	-Sufficient size -Various rooms -Can attach multi units -Private T&K	-Does not consider SN	-Durable -RES, Biogas toilet and rainwater harvesting -Use of foam insulation	-Use of steel		Unaffordable	[69,70]
12	Weaving a home	-Culturally acceptable	-Short-term solution -No private T&K	-RES and rainwater harvesting	-Use of plastic		'Unknown cost'	[71,72]

T—Toilet/K—Kitchen/SN—Social Needs/M—Materials/L—Local/G—Global/RES—Renewable Energy Sources/TC—Thermal Comfort/P—possible/Pe—Perlite/CS—Corrugated Sheets.

Table 2. Existing solutions sustainability comparison.

	Shelter Solution	Social Sustainability		Environmental Sustainability		Economic Sustainability		Ref
		Pros (+)	Cons (−)	Pros (+)	Cons (−)	Pros (+)	Cons (−)	
1	Refugee Housing Unit	-Easily deployed -Moveable	-Does not consider SN -One room -Small size -No private T&K	-Small solar panel	-Short lifespan -Use of steel -Use of plastic	Below average		[73,74]
2	Bangladesh 2007	-Expandable -LM	-Small size -T&K (unknown)	-Wind protection -Flooding protection -Self-built -LM- ex. Bamboo	-Permanent base -CS roof -Use of concrete		Above average (Material Costs)	[75]
3	Kenya-Dadaab 2009	-Culturally acceptable -Women participation	-Small size -T&K (unknown)	-Self-made mud blocks -Durable	-CS roof -Limited by MA -Unplanned excavation	Below average (Material Costs)		[75]
4	Haiti 2010	-Sufficient size -Outdoor porch -Traditional techniques -Flexible & Accessible	-Internationally procured materials -T&K (unknown)	-Use of Traditional M- wood/ mud (p) -Passive C -Durable	-Use of concrete -Corrugated bitumen roofing		Above average (Material Costs)	[20]
5	Philippines 2011	-Traditional techniques -Easily deployed -LM	-Small size -T&K (unknown)	-Durable -Use of wood -LM	-CS roof -Use of concrete	Below average (Material Costs)		[20]
6	Ethiopia 2011	-Various sizes -Built by refugees -LM -Separate private T	-Does not consider SN -No planned animals' shelters -K (unknown)	-Passive C&H -Use of LM, such as bamboo, grass and mud	-Sourcing issues -Transporting issues -Seasonal materials	Below average (Material Costs)		[76]
7	Madagascar 2012	-Culturally acceptable -Easily deployed -LM	-Small size -T&K (unknown)	-Use of wood -Thatch roof (P) -LM	-Unconsidered LM -CS roof (P)	Below average (Material Costs)		[76]
8	Fiji 2012	-Sufficient size	-T&K (unknown)	-Prefab elements -Withstands cyclones -Raised earth floor	-Use of plastic -CS roof		Above average (Material Costs)	[77]
9	Myanmar 2012	-LM	-Small size -Collective shelter -Does not consider SN -T&K (unknown)	-LM	-Seasonal materials	Below average (Material Costs)		[77]
10	Philippines 2012	-Various sizes -Separate T -LM	-Small size -K (unknown)	-Use of salvaged M -Use of fallen trees -LM	-Lack of salvaged M	Below average (Material Costs)		[77]
11	Jordan 2013	-Easily deployed	-Does not consider SN -One room -Small size. -No private T&K	-Use of foam	-Short lifespan -Use of steel & CS -Unsealed walls		Above average (Material Costs)	[31,77,78]
12	Iraq 2015-2016	-Locally procured GM -Divided interior -Private T&K	-Small size -Not flexible	-Wood & fiberglass -PU insulation -Durable	-Use of steel		Within existing range (Material Costs)	[27]

T—Toilet/K—Kitchen/SN—Social Needs/M—Materials/L—Local/ G—Global/C—Cooling/H—Heating/P—Possible/CS—Corrugated Sheets/Al—Aluminum/A—Availability.

The average material costs of the 12 examined 'existing solutions' which were funded by UNHCR or IFRC was calculated in this research as $1300. The maximum material costs of the 12 cases was $5500 for the Iraq 2015–2016 project [27]. These two costs along with their average ($3400) formed the basis for describing the costs in Table 3 that were used in Tables 1 and 2.

Table 3. Cost classification adopted in this research.

Materials Costs	<$1300	$1300–$3400	$3400–$5500	>$5500
Description	Below average	Above average	Within existing range	Unaffordable

4. Results

As explained in the literature review section, the shelter terminology is not always agreed upon between NGOs and academia. In addition, designs and solutions with assigned shelter types do not always meet the specifications of that type. Most of the 'novel designs' were considered as global shelters or as a one-size-fits-all solution, which is recognized as an inadequate approach as it neglects the social context and cultural needs [19,78]. Conversely, 'existing solutions' were designed for a specific case (full details given in the supplementary material). With respect to portability, most of the 'existing solutions' were fixed even though some of them were originally designed to be relocatable, but changes occurred during implementation such as the case of Jordan in 2013 (Azraq camp T-shelters). All of the 'novel designs' were transportable and 92% of them can be easily deconstructed; in most cases, they were flat packed (76%), but other techniques were also used such as being stackable, foldable, or able to be disassembled into smaller parts (details can be found in the Supplementary Material). Each sustainability dimension is discussed in detail in the following sections.

4.1. Social Sustainability

Defining social sustainability is challenging as the field is still emerging [79], while at the same time is vague and impossible to be limited in one definition [80]. However, in this paper, the adopted definition is the preservation of the existing social systems, where the social challenges and concerns are being addressed by considering the history, traditions, dialogue, equity, and participation [79]. The concept of social sustainability sometimes overlaps with the other two dimensions of sustainability, i.e., environmental and economic. However, each dimension tackles the overlapped element in different ways. A clear example is the materials, where the use of local materials fulfils a social need due to its familiarity, besides the fulfillment of the environmental and economic elements. Among the studied cases, the most commonly identified social pros for shelters that belong to both types (novel designs and existing solutions) were: the short time needed to assemble the shelters by a minimum number of workers, the ease of deployment that allows unskilled beneficiaries to take part in the construction, the use of local and locally available materials, the flexibility and expansion possibilities, having various shelter sizes to meet the needs of different household compositions, and having an interior layout that is divided into needed functions. In addition, some of the 'existing solutions' were more respectful by adopting local building techniques, having outdoor private areas, or using shelter types that are acceptable to users (i.e., familiar and used within their culture), such as Haiti 2010 [20] and Philippines 2012 [81].

Conversely, the most common cons under the social dimension are the one-room approach that is more emphasized in the 'novel designs' with 58% of the studied cases. In addition, the lack of private toilets and private kitchens is a major drawback, as only 8% of the 'novel designs' and 17% of the 'existing shelters' are known to have a private toilet, while 17% of the 'novel designs' and 8% of the 'exisiting shelters' have a private kitchen. The Refugee Housing Unit designed by IKEA shown in Figure 1a, is amongst the many shelter examples of the one-room design that also lacks private facilities [78]. The small or insufficient shelter area (compared to the number of residents and/or their needs) is another common con amongst both clusters. The Tentative Concept post-disaster shelter which is shown in Figure 1b is an example with its 8 m^2 overall area [67].

Figure 1. Shelters with social inadequacy: (**a**) Refugee Housing Unit [78], (**b**) Tentative Concept post-disaster shelter [67], (**c**) Myanmar 2012 temporary collective shelter, photo by UNHCR [81], (**d**) Ethiopia 2011 semi-permanent shelter, photo by Demissew Bizuwerk/IOM Ethiopia [20].

Other common cons were the total dependency on the availability of local materials in the location or total dependency on global materials, and proposing short-term solutions. Building collective shelters instead of private shelters was also apparent in the 'existing solutions' such as the Myanmar 2012 shelter that hosts eight families (Figure 1c) [81]. Providing a shelter design that is familiar to the host community but not to the users is another issue that was highlighted in the case of Ethiopia 2011. The Tukul shelters that were given to the refugees were familiar in the Ethiopian culture but not for the Sudanese who lived in them (Figure 1d). In the same case, another social inadequacy appeared when there was no space to shelter the livestock brought by the Sudanese refugees from their homeland [76]. Recent evidence has also shown that elements of social sustainability should go beyond the human sphere and take into account livestock and domestic animals because of their important role in certain cultures [82].

Identifying the best practices amongst the existing shelter projects and understanding the cultural influence on the spatial needs, would shorten the distance between the decision makers and the shelter users. When possible, engaging the users in the early stages of designing the shelter projects would ensure the social suitability of the final output.

4.2. Environmental Sustainability

The pros within the environmental dimension are related to the use of local materials such as wood, thatch and earth, the reusability and durability of the shelter, using passive cooling and heating techniques, the ability to collect rainwater and the provision of electricity through solar panels. In qualifying the environmental sustainability, the relative environmental consequences of different elements need to be borne in mind (e.g., the embodied energy in the materials and transport to site, the operational energy demand, the environmental consequences of rainwater harvesting).

The 'existing solutions' showed some good practice that was not seen in the 'novel designs', such as raising the shelter over a plinth in flood-prone areas, and having construction details that can withstand severe wind loads in cyclone-prone areas. It can be argued that 'existing solutions' have gone through a number of trial and error phases, which led to their better suitability to conditions of and application in the field. 'Novel designs' should certainly learn from the iterative experience of 'existing solutions'. Other positive approaches were the use of salvaged materials, using materials which were produced on-site by the beneficiaries, and adopting some traditional construction techniques, such as Clissage (a mix of lime and earth binding the filling of a wooden structure) and Amakan (woven bamboo wall cladding).

Conversely, the poor practice included using carbon-intensive materials such as concrete, plastic, steel, nylon and aluminium. The U-dome shelter shown in Figure 2a is an example of a shelter made of such materials. It consists of corrugated polypropylene panels, which are connected by nylon fasteners [52,83]. The Concrete Canvas shelter (Figure 2b) is another example where concrete was used for the outer skin [57]. There are also examples of using carbon-intensive materials amongst the 'existing solutions' case studies, such as the T-shelters provided for the Syrian refugees in Jordan. This shelter, which can be seen in Figure 2c [84], is made of an interlocking steel structure and covered with a double layer of Inverted Box Rib (IBR) cladding separated by an aluminium and foam insulation [81].

(**a**)   (**b**)

(**c**)   (**d**)

Figure 2. Shelters with environmental inadequacy: (**a**) U-dome transitional shelter (designboom, 2018), (**b**) Concrete Canvas shelter [57], (**c**) Jordan 2013 T-shelters, photo by Robert Neufeld (World Vision, 2014), (**d**) Kenya-Dadaab 2009 core shelter, photo by Jake Zarins [75].

In some cases, the demand for natural materials exceeded their availability, and this happened in the 'Kenya-Dadaab 2009' project (Figure 2d) where the insufficient availability of mud and shortage of water have limited the number of built shelters. Transporting mud from distant locations was proposed for future projects, though it turns the use of mud blocks into a less sustainable option for the rest of the shelters. The unplanned excavation of mud resulted in holes, which in the rainy seasons were transformed into refuse pits or mosquito-breeding sites [75]. In the 'Ethiopia 2011' project, a similar challenge occurred when there were difficulties in sourcing the mud and grass used for thatching [76]. In the same way, in the 'Myanmar 2012' project, the bamboo was not in season and a lower quality alternative was used [81].

4.3. Economic Sustainability

About 83% of the 'novel designs' analysed, have costs that are significantly higher than the 'existing solutions' or with unknown cost. Figure 3a shows the Hex House, a shelter designed by Architects for Society. In the Dezeen online magazine, the cost per unit was denoted as $15,000–$20,000 [70], while in the Hex House website, it is shown as $55,000–$60,000 [69]. Another novel shelter design with a high cost is the Rapid Deployment Module (Figure 3b) [65], with unit costs around $15,000–$18,000 [64]. On the contrary, 67% of the 'existing solutions' had modest costs (equal to or less than $1300). However, there were cases with costs that are considered high compared to other shelters such as Fiji 2012, which is shown in Figure 3c. Its material costs were $1800 and the overall project cost per shelter was $2900. The main reason for its high cost is likely to be Fiji's remote location, which increased the materials' transportation cost and therefore the overall cost [81]. Another 'existing solution' with a high cost is the Iraq 2015–2016 transitional shelter shown in Figure 3d. The shelter had material costs of $5500 and a project cost per household of $9621. The initial costs for establishing the sites were high and the political situation and the higher shelter standards also contributed to the high cost [27].

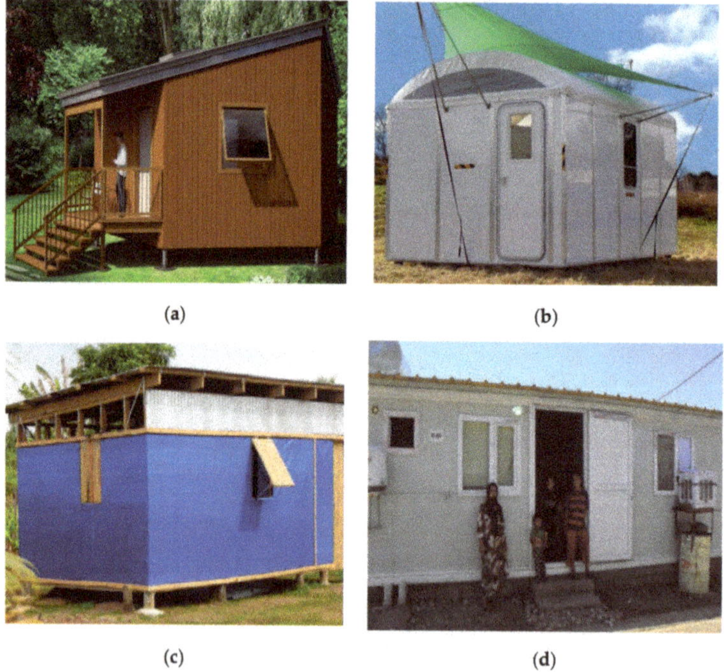

(a)

(b)

(c)

(d)

Figure 3. Shelters with economical inadequacy: (**a**) Hex House shelter [69], (**b**) Rapid Deployment Module, photo courtesy of RDM and Fast Company [65], (**c**) Fiji 2012 transitional shelter, photo by Habitat for Humanity Fiji [81], (**d**) Iraq 2015–2016 transitional shelter, photo by Alan Miran [27].

5. Discussion

The review of academic literature around PDPC sheltering and this research on 'existing solutions' and 'novel designs' allow us to understand best practice and common pitfalls across the three main sustainability dimensions. The main lesson learnt from previous shelter cases in regard to the social dimension is that one room designs do not meet social needs. The possibility of adding an internal fabric division to a single room does not match the performance of a more substantial wall. One of the common challenges in both the 'novel designs' and the 'existing solutions' is that the toilet/shower and the kitchen are not considered during the design phase. Private facilities are a major need in many cultures and failing to provide them leads to social, health and psychological problems.

Adding those private facilities at a later stage usually results in a waste of time, requires additional resources and incurs higher costs. In addition, the size of the shelter should match the number of space users, their age and gender. Providing one size shelter does not respond to diverse family needs. The Sphere project [4] recommends a minimum covered area of 3.5 m^2 per person, and despite the fact that this number has no scientific basis [85], most designs do not even meet such minimum recommended area. The minimum acceptable covered space per person will differ between contexts and cultures. Using materials that are familiar or accepted within the residents' culture, as well as being maintainable, are important social elements to consider. The shelter should be adequately designed before being distributed as it should not depend on the individual ability to source additional materials. In addition, providing spaces for the residents' animals is an important need in cultures where animals play a significant role. The primary recommendation to fulfil the social sustainability aspect in any shelter design would be to consider the importance of engaging with the residents in the design from early stages. That would help in providing a more satisfactory shelter, which responds to their own cultural needs and at the same time enhances their sense of ownership of their shelters.

In the environmental dimension, it was noted that all renewable energy applications are positive additions to any shelter design. However, it should be considered that these renewable sources cannot be the only energy source as they generally depend on weather conditions, which are uncertain. In addition, those applications are only cost effective if considered over the long term, and in most cases the duration of the situation is unknown, and budget is limited. Using natural materials like wood, bamboo, thatch, mud and other bio-based or recyclable materials could reduce environmental impacts, but this can only be explicitly analyzed through, for instance, life cycle assessment (LCA) and evidence, rather than the designer's intuition, which frequently drives design choices. Self-made materials such as woven bamboo mats or mud blocks can save money, increase the residents' sense of ownership and be at the same time more sustainable. If seasonal materials are used in the design, then a planned alternative should be identified for cases when the need for shelter falls out of that season. In general, the use of local materials is preferable but prefabrication could in some cases save time, and provide the necessary thermal comfort. Whatever is the selected approach, designing a shelter that can withstand the local weather conditions is a priority, especially in areas prone to severe weather phenomena. Appropriate passive cooling and heating techniques are usually found in a region's vernacular architecture and traditional houses. These techniques are generally more sustainable and familiar to the people. Utilizing them can have positive impacts on both the social and environmental dimensions. There could also be, however, a tension between greater environmental sustainability and greater social sustainability. For instance, in Uganda, as soon as people can afford to, they will get rid of environmentally-friendly thatched roofs as they "harbor pests and disease and are high maintenance, and will upgrade to a metal of tiled roof" [86]. So much so, that the Pulse Lab Kampala is using roof types as an indicator to measure poverty [86].

Materials are also influenced by the temporary nature of most shelters. Permanent shelters are not allowed in most cases, especially after conflicts, where the status of the land is a major concern. The limitations on the approved materials unfortunately direct organizations toward unsustainable materials which do not guarantee adequate levels of protection, such as plastic sheeting and corrugated galvanized sheets. Therefore, a pre-planned sustainable option could be explored and considered for each region, in order to alleviate the time pressure due to the urgency of PDPC situations. The lifespan of shelters and options for their reusability/recyclability should be considered while evaluating alternative designs to have a more realistic understanding of their values.

Most of the 'novel designs' have unrealistically high cost that exceeds that which UNHCR, IFRC and their operational partners usually pay for shelters. This difference is clearly noted by comparing the cost of the 'novel designs' with the cost of the 'existing solutions'. Since some of the 'novel designs' have never been prototyped, not all of them had published cost. On the other hand, the 12 studied 'existing solutions' had published material costs (not the full cost), and they range from $380 for the Philippines 2012 [81] to $5500 for the Iraq 2015–2016 [27]. The average material costs for the 12 cases is $1300. There is no fixed preferred cost for shelters, but the calculated average material cost can give an indication for what is considered affordable for PDPC shelters. It is vital to understand that the goal of a reduced shelter cost is not only to save money but most importantly to help more people within a fixed budget. Usually the shelter project beneficiaries are much fewer than the affected people who need help. Therefore, the principal purpose is to give the best shelter quality at the lowest possible cost to help the maximum possible number of people in need.

6. Conclusions

In the humanitarian sector, there are no agreed terminologies regarding sheltering and the existing terms are usually misused, which can mislead researchers and policy makers. The classification of shelters depending on their historical application values the field implementation, as it differentiates between the shelters that were already used by beneficiaries in real PDPC scenarios (i.e., existing shelters), and the shelters that were designed for PDPC scenarios but were not implemented and used by the affected people (i.e., novel designs).

It is widely understood that a compromise is always necessary in designing shelters, specifically between cost, performance, durability, cultural appropriateness and building technologies [20]. Considering sustainability in this complex scenario, which is often further constrained by resources and time, is inevitably challenging. However, it has a vital role to play in the wider wellbeing of the displaced. Currently, both existing and novel shelter solutions as described in this article fail to adequately meet the users' needs and a rethinking is therefore necessary. Additionally, the global need to address the climate crisis, and the consequential social benefits of more sustainable designs, push for a more holistic view of shelter design, one that addresses all three pillars of sustainability. Table 4 illustrates a summary of considerations when dealing with PDPC shelters in both the design stage and choosing materials, against the three dimensions of sustainability.

The main limitation in this research paper is the dependence on the available documented information instead of testing the prototypes themselves or conducting field visits and surveys. Moreover, the lack of an agreed documentation form for the shelter projects, resulted in having information which is neither consistent nor harmonized and therefore limited the scope of the compared indicators. The information was cited as found in the references, which means that they are not verified, specifically the 'novel designs'. This limitation forced a degree of qualitative interpretation to classify the indicators into pros and cons. Another limitation is the wide sample of studied shelters that were not focused on the responses of a certain disaster or a geographic region, which could offer clearer and more specified criteria for designing PDPC shelters. Future research could target a narrower group of shelters based on disaster type, location of shelters, culture, method of building, and/or political situation. Future research could also seek other ways of sourcing information, such as lab tests and field visits, to be able to collect the necessary data to achieve quantitative results.

Acknowledging and identifying compromises is a way to clarify possible future amendments and to achieve more sustainable shelter designs in future. This is what this article has sought to achieve, to comprehensively review the 'novel designs' and 'existing solutions' for post-disaster and post-conflict sheltering and conclude possible consideration for future shelter designs. The review was based on the available academic literature, organization reports and desk-based research to assess sustainability dimensions of shelter, which proved to be very limited. Both shelter clusters have been evaluated against social, environmental, and economic sustainability, in an attempt to qualify pros and cons along each dimension. Rather than to prescribe future design efforts, the aim of this paper was intended to identify best practices and successful and unsuccessful elements of design. A clearer understanding of successful approaches and areas for improvement represents an important stepping-stone for future work in this area to accelerate the transition to sustainable shelter solutions.

Table 4. Recommended PDPC shelter considerations.

	Social Sustainability	Environmental Sustainability	Economic Sustainability
DESIGN PHASE	Engage the residents in the designing phase	Use renewable energy applications, when applicable	(No available information on the cost of the design phase)
	Include private facilities whenever required	Appropriate passive cooling and heating techniques could be adopted from the region's vernacular architecture	
	The size must match the number of residents, their age and gender requirements	Consider the life-span of the shelter- Make it reusable or recyclable	
	Include spaces for the animals when needed		
CHOOSING MATERIALS	Familiar or accepted within the residents' culture	Use natural materials and other bio-based or recyclable materials, when applicable	The material costs for the shelter is preferred to be around $1300
	Maintainable	Use self-made materials whenever possible	
	Available	The use of seasonal materials shall be accompanied with a planned alternative	The material costs shall not exceed the maximum amount of $5500
	Temporary whenever the status of the land is a concern	A mix between local materials and prefabrication could be useful depending on the case	

Supplementary Materials: The following are available online at http://www.mdpi.com/2071-1050/12/3/890/s1 and http://www.mdpi.com/2071-1050/12/3/890/s2, Table S1: Extended table—Novel designs sustainability comparison, Table S2: Extended table—Existing solutions sustainability comparison.

Author Contributions: F.P. had the original idea of investigating the subject within the context of the project above. L.A. carried out the data collection and wrote the first draft of the article. F.P., L.A., B.D., S.S., and P.G. all contributed to subsequent drafts which gave the manuscript its final form. All authors have read and agreed to the published version of the manuscript.

Funding: The research presented in this article was funded by the Royal Academy of Engineering, under their Frontiers of Engineering for Development Programme, Grant No. FoESFt5\100015.

Conflicts of Interest: The authors declare no conflict of interest.

References

1. UNHCR, Figures at a Glance. 2019. Available online: https://www.unhcr.org/figures-at-a-glance.html (accessed on 17 August 2019).
2. Giustiniani, F.Z. New hopes and challenges for the protection of IDPs in Africa: The Kampala Convention for the Protection and Assistance of Internally Displaced Persons in Africa. *Denver J. Int. Law Policy* **2011**, *39*, 347.
3. Shelter Centre, IOM. *Transitional Shelter Guidelines*; Shelter Centre: Geneva, Switzerland, 2012. [CrossRef]
4. Sphere Project. *Sphere Handbook: Humanitarian Charter and Minimum Standards in Disaster Response*; The Sphere Project: Geneva, Switzerland, 2011.
5. Fosas, D.; Albadra, D.; Natarajan, S.; Coley, D.A. Refugee housing through cyclic design. *Archit. Sci. Rev.* **2018**, *61*, 327–337. [CrossRef]
6. Johnson, C. Strategic planning for post-disaster temporary housing. *Disasters* **2007**, *31*, 435–458. [CrossRef] [PubMed]
7. Pomponi, F.; Moghayedi, A.; Alshawawreh, L.; D'Amico, B.; Windapo, A. Sustainability of post-disaster and post-conflict sheltering in Africa: What matters? *Sustain. Prod. Consum.* **2019**, *20*, 140–150. [CrossRef]
8. Ramboll. *Save the Children, Shelter Innovations Workshop*; Ramboll: Copenhagen, Denmark, 2017.
9. Albadra, D.; Coley, D.; Hart, J. Toward healthy housing for the displace. *J. Archit.* **2018**, *23*, 115–136. [CrossRef]
10. SWS WCED. World Commission on Environment and Development. *Our Common Future* **1987**, *17*, 1–91.
11. Potangaroa, R. Sustainability by Design: The Challenge of Shelter in Post Disaster Reconstruction. *Procedia Soc. Behav. Sci.* **2015**, *179*, 212–221. [CrossRef]
12. Mateus, R.; Bragança, L. Sustainability assessment and rating of buildings: Developing the methodology SBTool PT eH. *Build. Environ.* **2011**, *46*, 1962–1971. [CrossRef]

13. Hadafi, F.; Fallahi, A. Temporary housing respond to disasters in developing countries-case study: Iran-Ardabil and Lorestan Province Earthquakes. *World Acad. Sci. Eng. Technol.* **2010**, *4*, 1276–1282.

14. Chang, Y.; Wilkinson, S.; Potangaroa, R.; Seville, E. Resourcing challenges for post-disaster housing reconstruction: A comparative analysis. *Build. Res. Inf.* **2010**, *38*, 247–264. [CrossRef]

15. Ashdown, P.L. *Humanitarian Emergency Response Review*; Department for International Development: London, UK, 2011.

16. Davis, I. What have we learned from 40 years' experience of Disaster Shelter? *Environ. Hazards* **2011**, *10*, 193–212. [CrossRef]

17. UNDRO. *Shelter after Disaster: Guidelines for Assistance*; Folder United Nations Disaster Relief Organization (UNDRO): New York, NY, USA, 1982.

18. Quarantelli, E.L.L. Patterns of sheltering and housing in US disasters. *Prev. Manag. Int. J.* **2005**, *4*, 43–53. [CrossRef]

19. Barakat, S. Housing reconstruction after conflict and disaster. *Humanit. Pract. Netw. HPN* **2003**, *44*, 1–37.

20. IFRC. *Post-Disaster Shelter: Ten Designs*; International Federation of Red Cross and Red Crescent Societies (IFRC): Geneva, Switzerland, 2013.

21. Félix, D.; Branco, J.M.; Feio, A. Temporary housing after disasters: A state of the art survey. *Habitat Int.* **2013**, *40*, 136–141. [CrossRef]

22. Quaglia, C.P.; Dascanio, A.J.; Thrall, A.P. Bascule shelters: A novel erection strategy for origami-inspired deployable structures. *Eng. Struct.* **2014**, *75*, 276–287. [CrossRef]

23. Opdyke, A.; Javernick-Will, A.; Koschmann, M. A Comparative Analysis of Coordination, Participation, and Training in Post-Disaster Shelter Projects. *Sustainability* **2018**, *10*, 4241. [CrossRef]

24. Opdyke, A.; Javernick-Will, A.; Koschmann, M. Assessing the impact of household participation on satisfaction and safe design in humanitarian shelter projects. *Disasters* **2019**, *43*, 926–953. [CrossRef]

25. REACH. *Al Za'atari Refugee Camp Shelter Assessment*; Registration Evaluation Authorisation and Restriction of Chemicals (REACH): Basel, Switzerland, 2014.

26. REACH. *Azraq Camp Shelter Assessment*; Registration Evaluation Authorisation and Restriction of Chemicals (REACH): Basel, Switzerland, 2015.

27. Global Shelter Cluster. *Shelter Projects 2015-2016*; International Organization for Migration: Dhaka, Bangladesh, 2017.

28. El-Masri, S.; Kellett, P. Post-war reconstruction. Participatory approaches to rebuilding the damaged villages of Lebanon: A case study of al-Burjain. *Habitat Int.* **2001**, *25*, 535–557. [CrossRef]

29. Barakat, S.; Zyck, S.A. Housing Reconstruction as Socio-economic Recovery and State Building: Evidence from Southern Lebanon. *Hous. Stud.* **2011**, *26*, 133–154. [CrossRef]

30. Davidson, C.H.; Johnson, C.; Lizarralde, G.; Dikmen, N.; Sliwinski, A. Truths and myths about community participation in post-disaster housing projects. *Habitat Int.* **2007**, *31*, 100–115. [CrossRef]

31. Alshawawreh, L.; Smith, R.S.; Wood, J.B. Assessing the sheltering response in the Middle East: Studying Syrian camps in Jordan. *Int. J. Humanit. Soc. Sci.* **2017**, *11*, 2016–2022.

32. Sphere, Humanitarian Chater and Minimum Standards in Humanitarian Response. In The Sphere Handbook, 2018 ed. 2018. Available online: https://spherestandards.org/wp-content/uploads/Sphere-Handbook-2018-EN.pdf (accessed on 23 January 2020).

33. Escamilla, E.Z.; Habert, G. Global or local construction materials for post-disaster reconstruction? Sustainability assessment of twenty post-disaster shelter designs. *Build. Environ.* **2015**, *92*, 692–702. [CrossRef]

34. Celentano, G.; Escamilla, E.Z.; Göswein, V.; Habert, G. A matter of speed: The impact of material choice in post-disaster reconstruction. *Int. J. Disaster Risk Reduct.* **2018**, *34*, 34–44. [CrossRef]

35. Tumbeva, M.D.; Wang, Y.; Sowar, M.M.; Dascanio, A.J.; Thrall, A.P. Quilt pattern inspired engineering: Efficient manufacturing of shelter topologies. *Autom. Constr.* **2016**, *63*, 57–65. [CrossRef]

36. Shelterprojects. Shelterprojects.org. 2018. Available online: http://shelterprojects.org/ (accessed on 28 October 2018).

37. Sadok, W.; Angevin, F.; Bergez, J.-E.; Bockstaller, C.; Colomb, B.; Guichard, L.; Reau, R.; Messéan, A.; Doré, T. MASC, a qualitative multi-attribute decision model for ex ante assessment of the sustainability of cropping systems. *Agron. Sustain. Dev.* **2009**, *29*, 447–461. [CrossRef]

38. Sarriot, E.G.; Winch, P.J.; Ryan, L.J.; Edison, J.; Bowie, J.; Swedberg, E.; Welch, R. Qualitative research to make practical sense of sustainability in primary health care projects implemented by non-governmental organizations. *Int. J. Health Plan. Manag.* **2004**, *19*, 3–22. [CrossRef]

39. Ricart, S.; Ribas, A.; Pavón, D. Qualifying irrigation system sustainability by means of stakeholder perceptions and concerns: Lessons from the Segarra-Garrigues Canal, Spain. *Nat. Resour. Forum* **2016**, *40*, 77–90. [CrossRef]

40. Pieroni, M.D.; Pigosso, D.C.A.; McAloone, T.C. Sustainable Qualifying Criteria for Designing Circular Business Models. *Procedia CIRP* **2018**, *69*, 799–804. [CrossRef]

41. Samarakoon, S.M.S.M.K.; Gudmestad, O.T. Retaining the Sustainability of Oil and Gas Operations: Qualifying the Best Available Techniques. In *American Society of Mechanical Engineers Digital Collection*; American Society of Mechanical Engineers: New York, NY, USA, 2010; pp. 11–17. [CrossRef]

42. Janoušková, S.; Hák, T.; Moldan, B. Relevance—A neglected feature of sustainability indicators. In *Routledge Handbook of Sustainability Indicators*; Routledge: Abingdon, UK, 2018; pp. 477–493.

43. Matard, A.; Kuchai, N.; Allen, S.; Shepherd, P.; Adeyeye, K.; McCullen, N.; Southwood, E. An Analysis of the Embodied Energy and Embodied Carbon of Refugee Shelters Worldwide. *Int. J. Constr. Environ.* **2019**, *10*, 29–54. [CrossRef]

44. Hammond, G.P.; Jones, C.I. Embodied energy and carbon in construction materials. *Proc. ICE Energy* **2008**, *161*, 87–98. [CrossRef]

45. Goodland, R. The concept of environmental sustainability. *Annu. Rev. Ecol. Syst.* **1995**, *26*, 1–24. [CrossRef]

46. Maxim, L.D.; Niebo, R.; McConnell, E.E.; McConnell, E.E. Perlite toxicology and epidemiology—A review. *Inhal. Toxicol.* **2014**, *26*, 259–270. [CrossRef] [PubMed]

47. Gargett, C. Emergency Shelter | Conrad Gargett. 2018. Available online: http://www.conradgargett.com.au/project/emergency-shelter/ (accessed on 14 December 2018).

48. Furuto, A. Emergency Shelter Winning Design/Nic Gonsalves + Nic Martoo. 2013. Available online: https://www.archdaily.com/385002/emergency-shelter-winning-design-nic-gonsalves-nic-martoo (accessed on 14 December 2018).

49. FIBONACCISTONE, EXO Housing Shelter—Stackable Emergency Housing from Reaction Housing—Fibonacci Stone. 2018. Available online: http://www.fibonaccistone.com.au/exo-shelter-stackable-emergency-reaction-housing/ (accessed on 14 December 2018).

50. McDaniel, M. Where can I buy an Exo Housing Unit? 2017. Available online: http://michaelmcdaniel.com/blog/2017/11/18/where-can-i-buy-an-exo-housing-unit (accessed on 14 December 2018).

51. Kessler, S. If Reaction Housing Wants to Provide Disaster Relief, It'll Have to Sh. 2015. Available online: https://www.fastcompany.com/3042416/hotel-30 (accessed on 14 December 2018).

52. Designboom. World Shelters: U Dome. 2018. Available online: https://www.designboom.com/architecture/world-shelters-u-dome/ (accessed on 30 November 2018).

53. World Shelters. World Shelters: U-Dome. 2018. Available online: http://worldshelters.org/shelters (accessed on 15 December 2018).

54. World Shelters. U-Dome 200, Arcata, CA, USA. 2009. Available online: http://worldshelters (accessed on 16 December 2018).

55. World Shelters. World Shelters: Transhel. 2018. Available online: http://worldshelters.org/shelters/transitional-shelter (accessed on 16 December 2018).

56. World Shelters. Transitional Shelter: Localizations to Recovery. 2018. Available online: http://worldshelters.org/wp-content/uploads/2010/01/haiti-relief-transhel-proposaldoc (accessed on 16 December 2018).

57. Concrete Canvas. Concrete Canvas. 2018. Available online: https://www.concretecanvas.com/cc-shelters (accessed on 26 November 2018).

58. Howard, B.C. Behind the Viral Sensation: Concrete Canvas Goes Beyond Fast-Deploying Shelters, National Geographic Society Newsroom. 2013. Available online: https://blog.nationalgeographic.org/2013/03/20/behind-the-viral-sensation-concrete-canvas-goes-beyond-fast-deploying-shelters/ (accessed on 16 December 2018).

59. Archdaily. Liina Transitional Shelter/Aalto University Wood Program. 2018. Available online: https://www.archdaily.com/174909/liina-transitional-shelter-aalto-university-wood-program (accessed on 16 December 2018).

60. Meinhold, B. The Liina Transitional Modular Shelter Needs No Tools for Assembly. 2011. Available online: https://inhabitat.com/the-liina-transitional-modular-shelter-needs-no-tools-for-assembly/ (accessed on 16 December 2018).

61. I-BEAM. The Pallet House. 2018. Available online: http://www.i-beamdesign.com/new-york-humanitarian-projects-design/ (accessed on 16 December 2018).

62. Lifeshelter. Lifeshelter—Innovative Shelter Solutions for Refugees. 2018. Available online: https://www.lifeshelter.com/family/ (accessed on 16 December 2018).

63. Real Relief. Lifeshelter. 2018. Available online: http://www.realreliefway.com/en-us/life-saving-products/habitat/lifeshelter®/shelters (accessed on 16 December 2018).

64. VisibleGood. Shelter: The RDM. 2018. Available online: http://visible-good.com/shelters/ (accessed on 26 November 2018).

65. Maxey, K. Shelter Assembly in 25 Minutes. No Tools Required. 2013. Available online: https://www.engineering.com/3DPrinting/3DPrintingArticles/ArticleID/5606/Shelter-Assembly-in-25-Minutes-No-Tools-Required.aspx (accessed on 2 December 2018).

66. Williams, A. "Rapid Deployment Module" Shelter Assembles in 25 Minutes, no Tools Required. 2013. Available online: https://newatlas.com/rapid-deployment-module/27077/ (accessed on 16 December 2018).

67. Treggiden, K. Designnobis Proposes Pop-Up Disaster Relief Shelter for Earthquake-Affected Families. 2015. Available online: https://www.dezeen.com/2015/08/23/designnobis-pop-up-temporary-disaster-relief-shelter-earthquakes-tentative/ (accessed on 26 November 2018).

68. DESIGNNOBIS. Tentative. 2018. Available online: http://designnobis.com/index.php?r=site/product&id=191 (accessed on 16 December 2018).

69. Hex House. Hex House—Rapidly Deployable House. 2018. Available online: http://www.hex-house.com/ (accessed on 26 November 2018).

70. McKnight, J. Architects for Society Creates Low-Cost Hexagon Refugee Houses, Dezeen. 2016. Available online: https://www.dezeen.com/2016/04/14/architects-for-society-low-cost-hexagonal-shelter-housing-refugees-crisis-humanitarian-architecture/ (accessed on 26 November 2018).

71. Abeer Seikaly. Weaving a Home. 2019. Available online: http://www.abeerseikaly.com/weavinghome.php (accessed on 10 March 2019).

72. Douglass-Jaimes, D. Abeer Seikaly's Structural Fabric Shelters Weave Refugees' Lives Back Together. 2015. Available online: https://www.archdaily.com/778743/abeer-seikalys-structural-fabric-shelters-weave-refugees-lives-back-together (accessed on 16 December 2018).

73. Fairs, M. Ten Thousand IKEA Refugee Shelters left Unused over Fire Fears, United Nations admits. 2017. Available online: https://www.dezeen.com/2017/04/29/united-nations-admits-10000-ikea-better-shelter-refugees-mothballed-fire-fears/ (accessed on 13 December 2018).

74. Better Shelter. Better Shelter. 2018. Available online: http://www.bettershelter.org/rebuilding-iraq/ (accessed on 10 April 2018).

75. UN-HABITAT; IFRC. Shelter Projects 2009. 2010. Available online: http://shelterprojects.org/shelterprojects2009.html (accessed on 13 December 2018).

76. IFRC; UN-Habitat; UNHCR. Shelter Projects 2011–2012. 2013. Available online: http://shelterprojects.org/shelterprojects2011-2012.html (accessed on 16 December 2018).

77. IFRC; UN-HABITAT; UNHCR. Shelter Projects 2013–2014. 2014. Available online: http://shelterprojects.org/shelterprojects2013-2014.html (accessed on 16 December 2018).

78. UNHCR. Shelter Design Catalogue, Geneva, Switzerland. 2016. Available online: https://cms.emergency.unhcr.org/documents/11982/57181/Shelter+Design+Catalogue+January+2016/a891fdb2-4ef9-42d9-bf0f-c12002b3652e (accessed on 16 December 2018).

79. Corsini, L.; Moultrie, J. Design for Social Sustainability: Using Digital Fabrication in the Humanitarian and Development Sector. *Sustainability* **2019**, *11*, 3562. [CrossRef]

80. Missimer, M.; Robert, K.-H.; Broman, G. A strategic approach to social sustainability e Part 2: A principle-based definition. *J. Clean. Prod.* **2017**, *140*, 42–52. [CrossRef]

81. IFRC; UN–HABITAT; UNHCR. Shelter Projects 2015–2016. 2016. Available online: http://shelterprojects.org/shelterprojects2015-2016.html (accessed on 16 December 2018).

82. Alshawawreh, L. Sheltering animals in refugee camps. *Forced Migr. Rev.* **2018**, *58*, 76–78.

83. Engineering Review International, U-Dome 200, Arcata-USA. 2009. Available online: http://worldshelters.org/wp-content/uploads/2010/02/u-dome200-5mm-eng-report-2009-10-301.pdf (accessed on 25 February 2019).
84. World Vision. New Refugee Camp Opens in Jordan as Syrian Humanitarian Crisis Continues to Grow. 2014. Available online: https://www.worldvision.org/about-us/media-center/new-refugee-camp-opens-jordan-syrian-humanitarian-crisis-continues-grow (accessed on 2 December 2018).
85. Kennedy, J.; Parrack, C. The History of Three Point Five Square Metres. In *Shelter Projects 2011–2012*; International Organization for Migration: Dhaka, Bangladesh, 2013; pp. 109–111.
86. United Nations Global Pulse. Measuring Poverty with Machine Roof Counting. 2018. Available online: https://www.unglobalpulse.org/projects/measuring-poverty-machine-roof-counting (accessed on 25 February 2019).

MDPI

St. Alban-Anlage 66

4052 Basel

Switzerland

Tel. +41 61 683 77 34

Fax +41 61 302 89 18

www.mdpi.com

Sustainability Editorial Office

E-mail: sustainability@mdpi.com

www.mdpi.com/journal/sustainability